Two Trains Leave Paris

NUMBER PROBLEMS FOR WORD PEOPLE

Taylor Marie Frey
+
Mike Wesolowski

ILLUSTRATIONS BY PATRICK TORRES

ABRAMS IMAGE, NEW YORK

THIS BOOK IS DEDICATED TO

every character in a math textbook
riding a train, baking a cake, building a bookcase . . .

. . . and to the students who read those
problems and said, "What?"

CONTENTS

Foreword

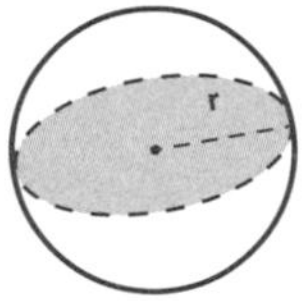

$$\frac{-b \pm \sqrt{b^2 - 4ac}}{2a}$$

In my own experience, Math courses through high school geometry consisted of three basic steps: (1) the teacher would show us how to do one or a few types of problems, (2) we would then do exercises imitating what the teacher showed us, and (3) we would take a test where we again did our best to imitate what the teacher showed us. This sounds a lot like learning to do a gymnastics routine, except without the thrill of flying through the air. In fact, I do not remember the process as being much fun at all. We were given no opportunity to experiment with the ideas involved (nor told there were any ideas), or to try out alternative ways of doing problems that would enable us to think about Math as a kind of play. Fortunately, this situation changes as one advances into calculus and beyond, or else no one would go into Math except people with a particularly imitative personality type. However, relatively few of us have the opportunity to get far enough in Math to regard it as a creative endeavor, and I fear many are left with a lasting impression of mind-numbing exercises and feelings of inadequacy.

Fortunately, teaching methods are changing, and there are more and more resources out there to help people enjoy Math. This book seems to me an excellent place to start if you want to have fun learning some Math and do not have a creative and innovative Math teacher at your beck and call. It is also an excellent choice if you need to learn some Math for a standardized test, and years of Math courses with anally retentive teachers dominate your impression of Mathematics.

This book is an apt selection of topics including Algebra, Geometry, Trigonometry, Calculus, Probability, and Statistics, and concludes with a chapter on

wedding planning, which is an underrated topic in Mathematics pedagogy. It is unlikely to teach you everything you need to know to get a PhD, but it will give you an idea of how to enjoy learning some important topics in Math, an experience that is valuable for multiple reasons. First, when you understand Math well, you learn that basically the same ideas are used over and over again, just in different settings. If you learn to play with these ideas, as this book encourages you to do, it will be much easier to think about how to use them in other contexts. Second, ideas you learn in an entertaining way will be much easier to remember when you need them. (I think you'll find that the approach to learning here will also help you in many other ways, perhaps including your next relationship crisis.)

Every year or two, I teach a course at Notre Dame where I do a version of the "dress for success" problem from Chapter Five. I try to make it humorous by mocking my poor grasp of color coordination, but my jokes don't work nearly as well as those of the authors, so I'll use the version in this book in my class going forward. If a few students happen to buy this book as a result, my motivation will have been purely educational, and not at all because my name is featured (in very small type) on the cover. At this point, I'll end the foreword, since if you really want to have fun learning math, you should stop reading this drivel I've written.

Dr. Samuel Evens
Professor of Mathematics
University of Notre Dame

Letters to Students

Dear Students,

Welcome back to Math—home of parallelograms, "solve for *x*," and the knowledge required for eighty-six percent of modern professions. Are you feeling a calming sense of nostalgia yet?

Like your math books of yesteryear, this one asks you to use logic, cunning, and creativity to solve its problems. Why wouldn't you show Math some gratitude? After all, it's done everything for you.

For millennia, Mathematics has served as the foundation of human society. From the development of early economic concepts, to proving the earth's revolution around the sun, to the building of great bridges, to the most advanced rocket science, every notable achievement by the human race can trace its origin to fundamental mathematical concepts.

Math is humanity's most famous and useful discovery. From elegant formulas such as $a^2 + b^2 = c^2$, to universal constants such as π & e, to the infinite uses of Trigonometry, nature has graced us with gifts that are incredibly complex and mysterious, yet still entirely clear and simple.

O, what could be more perfect?!

This author hopes that you, the student, upon closer examination of the problems in this book, will come away with a greater appreciation for Mathematics and its many applications to human existence. We are not perfect, but Math is. Let us embrace it, learn to appreciate it, and use it to work out our problems.

Respectfully,
Mike Wesolowski

Dear Students,

Welcome back to Math—home of parallelograms, "solve for *x*," and the knowledge required for five percent of day-to-day life. Are you having a panic attack yet?

Unlike your math books of yesteryear, this one doesn't ask you to leave your emotions at the door. Why would you spare Math from your feelings? What has it ever done for you?

For a long, unspecific amount of time, Math has tried in vain to make meaning out of chaos. From building things to breaking things, all notable human actions can be "conveniently" linked to some sort of Math. Math prides itself on teaching students the formula to build a thing without asking them to build it, or to understand its purpose on an emotional or personal level.

Math is the infamous enemy of students, taunting them with cryptic formulas, little numbers next to big numbers; sometimes numbers on top of each other; sometimes lines, squiggly lines, straight lines, arrows, sometimes they point up, sometimes they go to the side, sometimes they're alligators?

O, what could be more confusing?!

This author hopes these problems give you some peace if you struggled with Math. Maybe it's time Math struggled to understand you. We are not perfect; Math is not perfect. So let us use each other, like strangers in the night, to work out our problems.

Respectfully,
Taylor Marie Frey

How to Use This Book

This book can be used to solve math problems, or not.

You can also use it in the following ways:

- To kill a spider
- As a coaster
- To fix a table leg
- As a present for your brother's girlfriend on her birthday (you don't know her that well)
- To take on the train, find someone else reading it, and fall in love
- To share with your math tutor
- To share with your therapist
- Or maybe both at the same time?

Please be prepared with the following:

- Pen/pencil/eraser/Wite-Out (not for sniffing . . .)
- Water
- Coffee, or any socially accepted addictive substance
- Calculator (the one on your phone is fine—joke's on that teacher who said, "You better learn how to do math without a calculator; you won't have one with you everywhere you go . . . ")
- The ability to self-reflect

See you in the book!

Chapter Zero

0

Zero: A number that has sparked an incredible amount of debate around something that literally means nothing. Zero is the absence of a measurable quantity: life before Math, Adam before Eve, Rama Lama before Ding Dong. We can't directly credit anyone with the invention of Zero, but it can be seen in the archives of ancient societies across the world, from the Maya to the Babylonians. Like dragons. And human sacrifice.

Zero is your safe place. If you leave Zero, you will find problems: math problems, emotional problems, existential problems . . . you name it. Moving away from Zero will complicate things, just like *your* life got more complicated when you left your mother's womb. Once you've had one, Zero will never be enough, and once you're at negative one, Zero will always seem like more than it is.

Do you wish to proceed past the mathematical womb that is this chapter? Please think deeply about this. Maybe you're fine exactly where you are, and don't need to think about the problems still to come, like Algebra, Geometry, Taxes, and Marriage. Feel free to stay in this chapter for a while, count on your fingers, go for a walk in the park, or sit and meditate on the concept of Zero.

If you do seek to complete the problems that lie within these pages, reading this chapter will give you a few tools to use. Here you will find practice problems that contain only the friendliest forms of Math: Addition, Subtraction, Multiplication, Division, and a few others.

Or just skip this section and start with the hard stuff right away. See if we care that you decided to skip a chapter we spent all this time on because you wanted to be a big shot.

PEMDAS

This shall be your lantern through the darkness. It is the **order of operations**, which tells you how to simplify a long expression like: $9 \div 3 + 2(5 - 3)^3 - 19$

No matter how complicated the situation gets, PEMDAS will always be the order of operations.

In some schools, students are taught a trick to remember PEMDAS, something like "Please Excuse My Dear Aunt Sally," which, like most American inventions, was designed to be both convenient and sexist. *There goes Aunt Sally again, speaking her mind. Please excuse her. And don't get too close! The doctor said she has "hysteria."*

PARENTHESES

That's these: () or [], (4) or [4]

Is the 4 being hugged? Or is it trapped in a little room?

EXPONENTS

Repeated multiplication of a certain amount, such as:

$$5^3 = 5 * 5 * 5 = 125$$

$$7^6 = 7 * 7 * 7 * 7 * 7 * 7 = 117{,}649$$

Exponents often result in large numbers,
but they are notated with tiny numbers.

MULTIPLICATION / DIVISION

Listed below are the many different ways to notate multiplication:

$$3 \times 4 \text{ or } 3 \cdot 4 \text{ or } 3 * 4 \text{ or } 3(4)$$

And division:

$$8 \div 2 \text{ or } 8/2 \text{ or } \frac{8}{2}$$

Why so many symbols, you ask?

Well, a long time ago, some guys didn't agree, so they just decided to never agree. They're dead now, but their traditions live on in boardrooms and houses of Congress and Twitter.

ADDITION / SUBTRACTION
These guys: + –

Seen on batteries, pregnancy tests, and our report cards from high school Math:
Mike: A+
Taylor Marie: D–

YOUR TURN:

Simplify $9 \div 3 + 2\,(5 - 3)^3 - 19$

First, do whatever is inside the **PARENTHESES** . . .

$$9 \div 3 + 2(2)^3 - 19$$

Now the **EXPONENT** . . .

$$9 \div 3 + 2(8) - 19$$

Now the **MULTIPLICATION** and **DIVISION** . . .

$$3 + 16 - 19$$

And finally, the **ADDITION** and **SUBTRACTION**.

$$0$$

That's the name of this chapter.

Fractions, Decimals, and Percentages

Since people have an insatiable need to compare themselves to other things, these are used all the time. Fractions, decimals, and percentages are all different ways to express *parts of a whole.*

Example

1 out of 2 can be written as

$\frac{1}{2}$

.5

50%

If they all mean the same thing, why don't we just use one way to represent it? Great question. Here's a similar question: Why does Maroon 5 record an entire album when every song is the same? And why were they spending all their change on you at a payphone in 2012? Some questions don't have answers—it's best that you accept this early on.

Ready to try?

Go Ahead, Try

#1. Simplify the following expression. Show your work.

$$1 - 3\,(2 - 1)^7 + 8 \div 4$$

Answer: Did you get Zero again? That's the name of this chapter.

#2. You have 5 fingers on your left hand, and 5 fingers on your right hand. How many fingers do you have?

Answer: That's right, 10 fingers.
(By the way, different numbers can add up to the same sum. Let's say you've got 4 fingers on your left hand, and 6 fingers over on your right hand . . .)

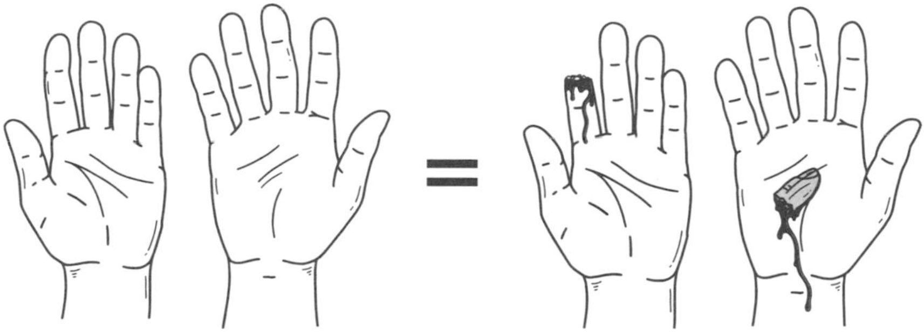

Mathematically speaking, these two images are equivalent. Do you agree?

#3. The first thing Taylor Marie ever wanted to be was an astronaut. In first grade, when she got her history book, she ripped out the pages about the moon landing and put them up on her wall. Her book was 475 pages long; she ripped out 6 pages. How many pages are left?

Answer: 475 − 6 = 469

The next day, her teacher asked the class, "What do *you* want to be when you grow up?" Taylor Marie said, "An astronaut!" The teacher said, "Well, an astronaut has to go to school for a long time and be very smart, and most astronauts are men who are good at Math." So Taylor Marie came home with one less dream than she went to school with. Subtraction.

#4. Mike has decided to multiply, and calls his math teacher to tell him he's expecting his first child. "Congratulations!" or "I'm so sorry," says the math teacher. "Did you know that someone in the strange business of quantifying human lives once estimated the cost of raising a child through age 17 to be $233,610?" *Uh-oh,* thinks Mike—the mother just called: She's expecting triplets. What are Mike's expected child-rearing expenses?

Answer: 233,610 * 3 = 700,830
More, if the children live beyond 17.

#5. You very much enjoyed ruling over the Kingdom of Targaryan, or whatever it is. You passed some good laws, fed the citizens, and for the most part kept the peace. Now the time has come for you to die and be reborn as a ladybug. Your final days are spent surrounded by your family, who sit by your side, singing songs and passive-aggressively arguing over your will. Hmmmm, seems like everyone wants a piece. In the end, you decide to dissolve your Kingdom of Targaryan, or whatever it is, and let the people govern themselves. The kingdom spans 9,000 acres, and you want to divide the land evenly among 1,500 households. How much land does each household get?

Answer: $9000 \div 1500 = 6$ acres per household.
Well played—as everybody knows, all great political leaders are skilled in Math. Good work! You may die now.

#6. You took a math test and got 60 answers correct out of 100 total answers. How can you express your score as a percent, decimal, fraction, and person?

Answers:
Percent: 60%
Decimal: .6
Fraction: 60/100 (or $\frac{6}{10}$ or $\frac{3}{5}$)
Person: act out socially, experience change in appetite, listen to Maroon 5

#7. Simplify the following:

A) $\frac{2}{4}$

B) $\frac{8}{6}$

C) Health care

D) Humanity's relationship to the sun

Answers:

A) $\frac{1}{2}$

B) $1\frac{1}{3}$ or $\frac{4}{3}$

C) Provide as much health care as possible to as many people as possible at as little cost as possible.

D) It is bigger, stronger, hotter, and brighter than us. Thank you, Sun, thank you every day for not killing us dead.

Alas, your moment has come. You hear a voice saying something like "Ma'am, you're 10 centimeters dilated, it's time to push." Now you, the fetus, must choose. You can remain in the proverbial womb that is this chapter, or you can follow the sweet, confusing sounds of Math.

Whatever you choose, proceed at your own risk.

WARNING: If you find yourself thinking *How could I, an imperfect human, ever understand Math, a perfect thing?!* then you are in danger of entering a toxic relationship. This book is not "approved by the scholastic community." This book is not "written by mathematicians." This book is not "724 pages." THIS BOOK IS NOT A TEST.

Reference Section

AREA / VOLUME FORMULAS

2-Dimensional Shapes: Area.
Expressed in "square units" such as "square inches" (in^2) or "square meters" (m^2).

Circle

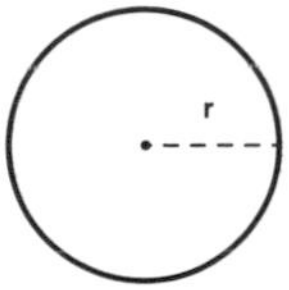

$A = \pi r^2$

$C = 2\pi r$

Rectangle

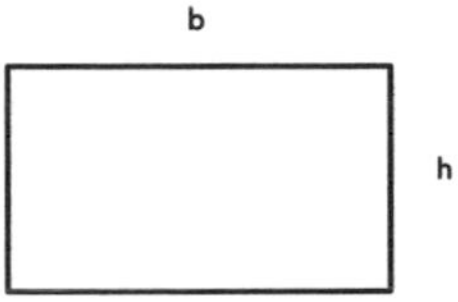

$A = bh$

Triangle

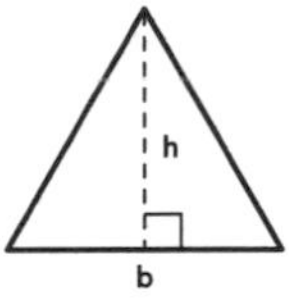

$A = \frac{1}{2} bh$

Gerrymandered Congressional District

A = Area, C = Circumference, r = Radius, b = Base, h = Height.
Gerrymandered Congressional District = unfairallelogram

3-Dimensional Shapes: Volume.
Expressed in "cubic units" such as "cubic inches" (in^3) or "cubic meters" (m^3).

Cylinder

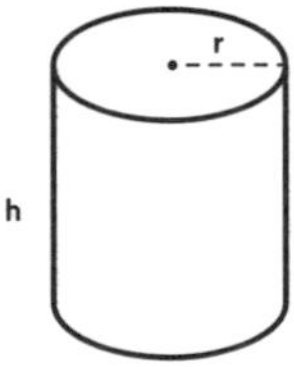

$V = \pi r^2 h$

Rectangular Prism

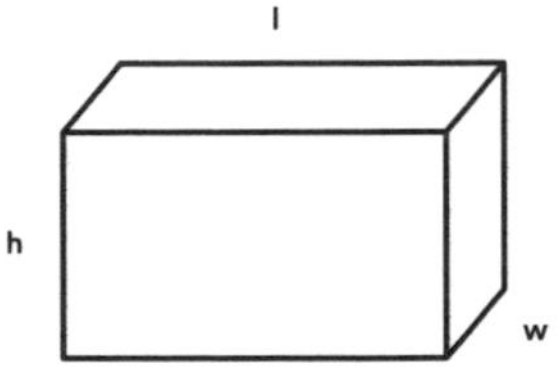

$V = lhw$

Sphere

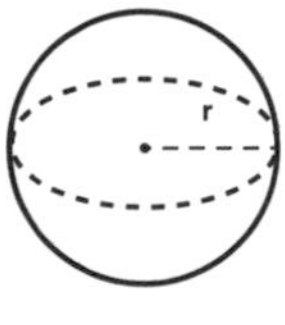

$SA = 4\pi r^2$

$V = \frac{4}{3} \pi r^3$

The Eye of God

V = 0 (silent)

V = Volume, SA = Surface Area, l = Length, w = Width

GEOMETRY

ANGLES

Acute	Right	Obtuse
$0° < x < 90°$	$x = 90°$	$90° < x < 180°$

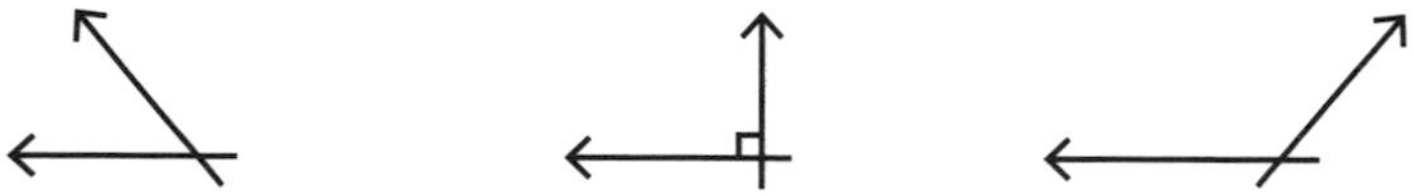

SPECIAL RIGHT TRIANGLES

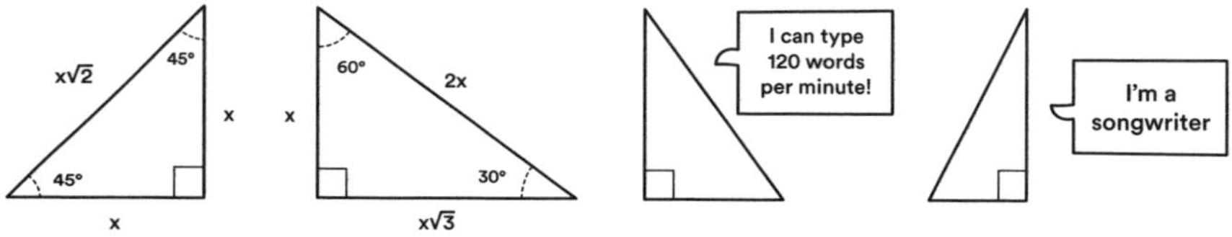

PYTHAGOREAN THEOREM

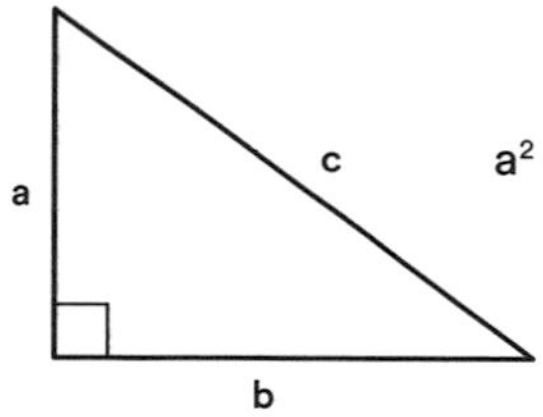

$a^2 + b^2 = c^2$

MORE GEOMETRY FACTS

- A circle's arc is 360°.
- A triangle's three angles in total are 180°.
- A right angle is 90°.
- An oft-forgot boy band is 98°.

THE QUADRATIC FORMULA

For an equation in this form:

$$ax^2 + bx^2 + c = 0$$

$$x = \frac{-b \pm \sqrt{b^2 - 4ac}}{2a}$$

CONVERSION CHART

1 inch (″)	=	2.54 centimeters (cm)
1 mile (mi)	=	5,280 feet (ft)
1 meter (m)	=	100 centimeters (cm)
1 gallon (gal)	=	4 quarts (Q)
1 minute (′)	=	60 seconds (″)
1 year (yr)	=	525,600 minutes (′)
time (🕑)	=	money ($)
money ($)	≠	happiness (☺)
time (🕑)	=	happiness? (😐)
corporations (🏭)	=	people (👪)
The love you take (<3)	=	The love you make (<3)

OTHER INFORMATION

WOMEN'S EARNINGS VS. MEN'S EARNINGS
1960 – 2015 (USA)

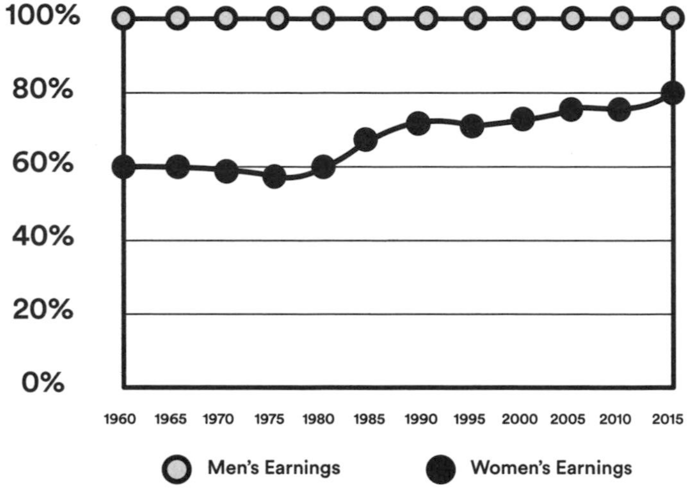

BLOOD ALCOHOL CONTENT

Drinks	Body Weight In Pounds							Influenced?
	100	120	140	160	180	200	220	
1	.05	.04	.03	.03	.03	.02	.02	Gettin' Loose!
2	.09	.08	.07	.06	.05	.05	.04	Impaired
3	.14	.11	.11	.09	.08	.07	.06	
4	.18	.15	.13	.11	.10	.09	.08	Legally Intoxicated
5	.23	.19	.16	.14	.13	.11	.10	
6	.27	.23	.19	.17	.15	.14	.12	
7	.32	.27	.23	.20	.87	.16	.14	
8	.36	.30	.26	.23	.20	.18	.17	
9	.41	.34	.29	.26	.23	.20	.19	
10	.45	.38	.32	.28	.25	.23	.21	Tiger Woods that one time

THE MOON

- Distance from Earth: 238,900 miles
- Gravity: 1.622 meters per second^2
- Inclination: 5.145° to the ecliptic
- Volume: 21,958,000,000 km^3 (2.19% of Earth's)
- Fun to howl at? Yes.

Chapter One

ALGEBRA

GETTING TO KNOW THE UNKNOWN

"Why do I even need to *know* Algebra?" you mumbled under your breath. "It's not going to help me in *real* life." Your mumbling distracted the teacher, but you were not wrong. Algebra is mostly just foundational work for advanced math topics like Trigonometry and Calculus. So it's great for preparing you to be a successful math student, but not great at preparing you to be a successful human.

But it may just be able to help you in one important aspect of your life: problem-solving. Algebra is all about breaking problems down into little pieces. No need to start with "Be happy!" when you can start with "Enjoy this nice piece of fruit."

Where did it come from?

Algebra's origins can be traced back to ancient Babylonia, but modern Algebra began to take shape around 800 CE, when polymath Muhammad ibn Musa Al-Khwarizmi wrote *The Compendious Book on Calculation by Completion and Balancing*. A riveting read—we can't wait for the film adaptation.

Since the word comes from the Arabic word *al-jabr*, meaning "the reunion of broken parts," you'd think Algebra would have more practical applications, like fixing broken glasses, mending your broken heart, or solving gender inequality. But clearly, that's not how it works.

How does it work?

Happy you asked. Algebra uses symbols to represent numbers in order to solve problems. This is where you would either "solve for x" or pretend you were sick so you could go to the nurse's office.

For example:

$$5x + 1 = 36$$
$$5x = 35$$
$$x = 7$$

The sad truth is: To solve for x, you must first separate it from everything it's attached to. You see, dear friend, Algebra is the loneliest branch of Mathematics—that's why you start learning it in middle school.

What's in this chapter?

You may think that because this is Chapter One, the problems here will be the easiest in the book, but don't be so sure. The problems in this chapter include the quadratic formula, the Pythagorean theorem, and an unquenchable desire to discover the unknown.

Some problems will be accompanied by instructions to solve; some will not. For the ones that don't include instructions, use what you learn from the ones that do! Or take your best guess. Or go get a bagel. Again—not a test.

Now, some of you math experts might have a look at this chapter and say something like "Hey! Not all of these problems are specifically algebraic in nature. There are elements of geometry and all sorts of other stuff in here!"

And we just want to say:

"Good for you."

#1. Two Trains

Two trains leave Paris, heading in different directions. Train A makes its first stop 60 kilometers due west. Train B makes its first stop 80 kilometers due north.

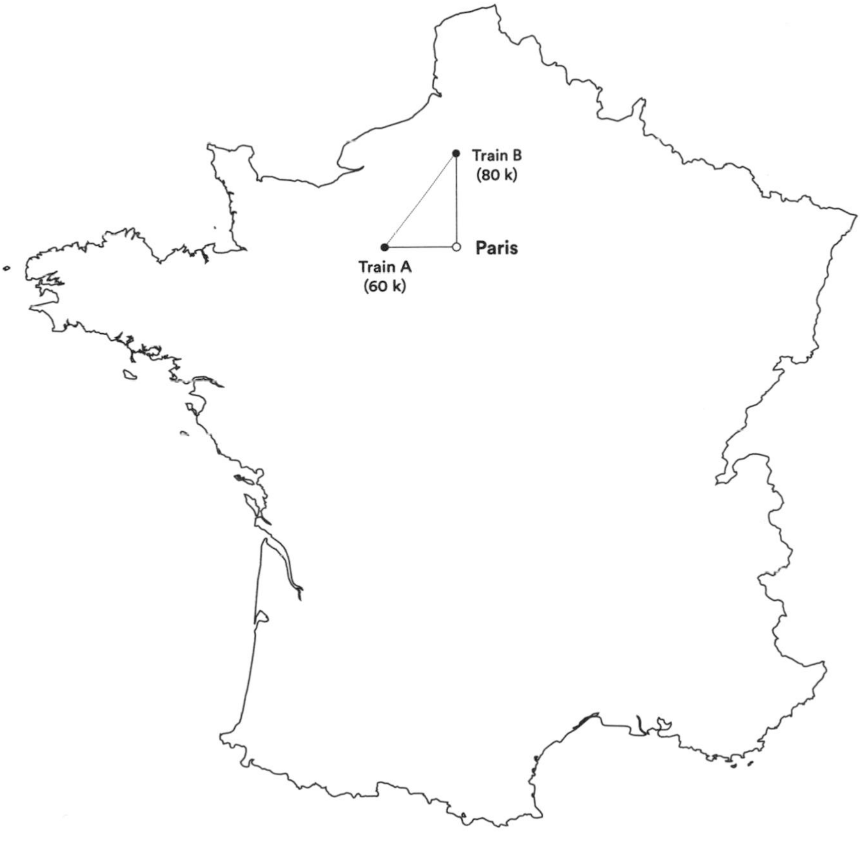

How much distance is between Natalya and Andy when they simultaneously look back toward the city of love, thinking, *Every end is a new beginning, and breaking up was for the best. No turning back now, this train only goes one way.*?

TO SOLVE — Two Trains

Use the Pythagorean theorem from the Reference Section. (Go ahead and look it up; no need to impress anyone here. Unless you bought this book to impress people. In which case . . . you should have bought a different book, like *Infinite Jest* or *Everybody Poops.*)

The Pythagorean theorem works like this: a and b represent the shorter sides of the triangle and c represents the hypotenuse—a Greek word meaning "the side across from the right angle."

Plug in the numbers and solve:

$$a^2 + b^2 = c^2$$
$$60^2 + 80^2 = c^2$$
$$3600 + 6400 = c^2$$
$$10{,}000 = c^2$$
$$100 = c$$

So Andy and Natalya are 100 kilometers apart *physically*, but in the exact same place *emotionally*.

Hint: For the last step, $10{,}000 = c^2$ to $100 = c$, you need to use the $\sqrt{}$ (or $\sqrt[2]{x}$) key on your calculator. Using your phone calculator and can't see $\sqrt{}$? Turn your phone to the side. Not sure what $\sqrt{}$ does? Put in some numbers like 4, 9, 16, and 25, and then hit $\sqrt{}$. The number breaks down into something smaller, like Natalya and Andy's breaking hearts. You're learning, they're learning, we're all learning. Let's learn $\sqrt{}$'s name. It's Tim.

Just kidding.

It's called the square root. The square root of 16 is 4 because $4 \times 4 = 16$. The square root of 100 is 10 because $10 \times 10 = 100$. But Tim, multiplied by himself? He's still just Tim. Let that be a vague lesson to you.

#2. What's My Age?

When asked his age, your math teacher, Mr. Newman, responds, "If you multiply my age by 4, then subtract 2, the answer is 110."

A) How old is he?

B) Why does Mr. Newman talk this way?

TO SOLVE — What's My Age?

A) First, identify the unknown.

In this case, the unknown is Mr. Newman's age.

Let's choose a letter to represent the unknown.

No, no, choose again. You're going to want to pick the most mysterious letter possible.

Did you pick x?

OK, good. Based on the words of the problem, write an equation like this:

$4x - 2 = 110$

Now add 2 to each side.

$4x = 112$

And finally, divide both sides by 4.

$x = 28$

B) He has a hard time opening up to people. Plus, he'd rather define numbers than let numbers define him.

#3. Evaluation

Evaluate the expressions below:

A) $3x^4 - 3x + 8y^2$ for $x = -2$ and $y = 9$

B) $4y^4 + x^2 + y^3$ for $x = 40$ and $y = -1$

C) Does she like you?

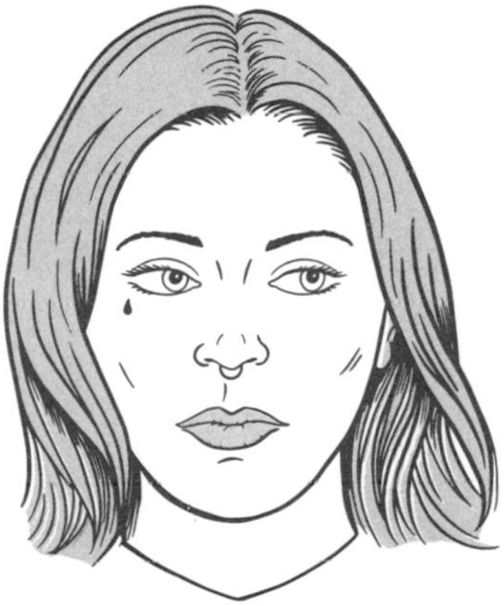

#4. Salary-ously, Guys?

At the hedge fund Ball, Shaft, & Associates, Bart's starting salary is $80,000, and he negotiates a $1,200 raise each year. Sonya's starting salary is $75,000, and she is awarded a $1,000 raise each year.

After how many years of working at BSA will Sonya's salary equal Bart's salary?

TO SOLVE — Salary-ously, Guys?

You can write equations for the two salaries. The number of years will be represented by x.

Bart = $80{,}000 + 1{,}200x$
Sonya = $75{,}000 + 1{,}000x$

Since we want to know when Bart and Sonya's salaries are equal, set the expressions equal to each other and solve for x:

$$80{,}000 + 1{,}200x = 75{,}000 + 1{,}000x$$
$$80{,}000 + 200x = 75{,}000$$
$$200x = -5{,}000$$
$$x = -25$$

OK, −25. So . . . 25 years ago? Like when they were kids?

Unfortunately, at this rate, their earnings will never be equal.

FYI: A number that starts higher and grows at a faster rate will *always* be larger than a number that starts lower and grows at a slower rate. If someone tells you otherwise, don't listen! Even if they're on TV and use phrases like "I understand. I have daughters."

#5. Theft!

Your rival, **McNiel**, has stolen your math book. You have a test tomorrow, so you're chasing **McNiel** down the street on your bike. **McNiel** has a 1000-foot head start and is traveling at 30 feet per second. You're coming in hot, literally pursuing knowledge at the breakneck pace of 31 feet per second.

A) How many seconds will it take you to catch **McNiel**?

B) How many minutes will it take you to catch **McNiel**?

C) Why did **McNiel** do this to you?

#6. Quest for the Crucial Information

The Half-Man Half-Shark lies slain, having come face to face with Cody's powerful sword. The Big Blue Crystal of Atlantis shimmers in the distance.

Nothing can stop Cody now—the secrets of the ancient city are within his grasp. He approaches the crystal, so big, so blue. He tries to pry it from its pedestal, but it's stuck. What could it be? He found the entry to Atlantis, crossed the scary bridge, scaled the slippery wall, saved the Princess, took her back to her mom's house, passed the crab's test for the pure heart, *and* battled the Half-Man Half-Shark. To the death. What else must he do?

Suddenly, mysterious symbols appear on the crystal. Symbols Cody doesn't understand, glistening in an ancient font. They read:

$$2x^2 - 4x - 6 = 0$$

Use the quadratic formula to solve for x.

What?

FACTS & OPINIONS (FROM MIKE)
STANDARDIZED TESTS

Standardized Testing is a method of assessing and comparing the abilities of a large group of people.

In education, the most popular standardized tests are the SAT and ACT, the scores of which are used on college applications.

Many states also administer yearly standardized tests to track the progress of students' learning. The results of these tests are also used to measure the effectiveness of individual teachers.

These tests are usually administered on specially scheduled days in large rooms, so that they can be given to a hundred students or more.

Standardized testing is useful for developing useful life skills for students entering the real world of adulthood: critical thinking, data analysis, and decision-making. In addition to the content, the practice of studying and taking the test teaches patience, discipline, and accountability.

It helps us track changes in our education system, and will continue to do so for the time being.

OPINIONS & FACTS (FROM TAYLOR MARIE)
STANDARDIZED TESTS

Standardized Testing is a very old, very scary torture practice developed in ancient China to assess the skills of a particular body of people.

Remember the SAT? The one where you just SAT there forever? Or the ACT? The one that ACTed like it was gonna be easy and then wasn't?

Some states force their students to take multiple tests every year, which eliminates hours and hours of valuable classroom time, and heightens the stress levels of teachers, who already work sixty to seventy hours a week.

In these tests, you'd sit in a school gym, or an auditorium, or a dark spooky dungeon, grip your #2 pencil in hand, speak with no one, and fight for your life.

Standardized testing teaches tons of useful life skills for students entering the real world of adulthood: filing your taxes, budgeting, conflict resolution, balancing and integrating technology, dealing with the cable company, and defense against the dark arts. Oh, sorry, I meant: sitting in rows, abusing ADHD medications, and filling in bubbles correctly (more complicated than it sounds).

It's ruining school, and must be stopped.

#7. Match Game

#1.	A number plus 5 is 12.	a.	$\frac{1}{3}(\textit{time}) = \textit{charm}$
#2.	A number divided by 2 is 11.	b.	$n - 18 = 40$
#3.	Showing up is half the battle.	c.	143
#4.	I love you / You are loved.	d.	$\textit{life} - x = \textit{death}$
#5.	Patience is a virtue.	e.	$x + 5 = 12$
#6.	18 is the difference between a number and 40.	f.	$\textit{showing up} = \frac{1}{2}(\textit{battle})$
#7.	The difference between life and death is unknown.	g.	$\frac{x}{2} = 11$
#8.	A bird in the hand is worth two in the bush.	h.	(🐦 hand) = (🐦🐦 bush)
#9.	Some things in life are worth waiting for.		
#10.	Third time's the charm.	i.	1 patience = 1 virtue
#11.	Third times the charm.	j.	$3 * \text{charm} = x$

#8. You're Like, My Best Friend

One afternoon, Natalya and Sonya, two adult women, meet at their local liquor store. Natalya, whose tolerance is high from her time in Paris, buys 3 bottles of wine that cost \$40 in total. Sonya, whose stress level is high from her job at Ball, Shaft, & Associates, buys 5 bottles of wine at a total cost of \$55.

Which expression best describes the combined contents of their checking accounts, each of which started the afternoon with x dollars in it, now?

A) $x - (3(40) + 5(55))$
B) $2x - 95$
C) $2x + 95$
D) "You only live once."

#9. The Space Between

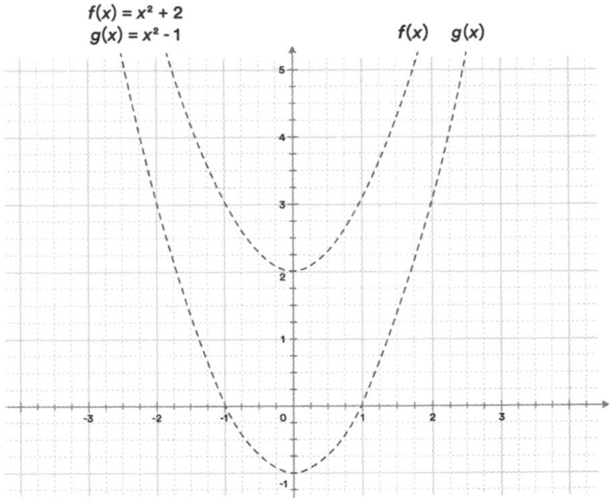

A) If the graphs of f and g intersect, find their intersection point.

B) If they do not intersect, how does it feel to be so close to something for so long but never be able to touch it?

#10. Which Came First: The Chicken or the Systematic Pollution of Our Natural Resources?

Anita, a fourth-generation farmer, has 18 chickens. Because she only has access to a heavily polluted water supply, some of these chickens have three legs. Her 18 chickens have a total of 40 legs.

How many mutant chickens does Anita have?

TO SOLVE — Which Came First: The Chicken or the Systematic Pollution of Our Natural Resources?

This time there are two unknowns.

Let x equal the number of regular chickens.
Let y equal the number of mutant chickens.

(We use y because it is the *second*-most mysterious letter.)

Set up two equations.

One for the number of chickens, and another for the number of legs:

$$x + y = 18 \qquad 2x + 3y = 40$$

$$x = 18 - y$$

$$2(18 - y) + 3y = 40$$

$$36 + y = 40$$

$$y = 4$$

$$x = 18 - 4$$

$$x = 14$$

She has 4 mutant chickens: Sarah, Pipsqueak, Clayton, and His Egg-sellency Haverford St. Wings. But she doesn't treat them any differently than she treats Donna or Eggbert or Sunny or Cooper or Waddles or Henry or Dr. Hayworth or Queenie or Sparky or Jennifer Aniston or Sir Clucksabunch or Mrs. Clucksabunch or Pikachu or Doug.

#11. Mr. Houdini Prescriptions

Natalya just finished her first day at a new part-time job: selling pot on behalf of Mr. Houdini Prescriptions Marijuana Delivery Service. She reaches into her repurposed camping backpack for her sales ledger when she remembers: She ripped it into tiny little pieces at her last stop. She didn't know what to do with her hands—the customer's dark hair reminded her of Andy.

She knows that she made $650 selling 2 strains—Kevin Wake'n'Bacon ($30 each) and Sativa Gomez ($55 each). At the 14 stops, each customer bought 1 kind, except for Norah, who grabbed 1 of each because her son is at Virtual Reality Camp this week.

How much of each sweet blend did Natalya sell today?

#12. Pascal's Numbers

These are the first ten rows of Pascal's triangle:

									1									
								1		1								
							1		2		1							
						1		3		3		1						
					1		4		6		4		1					
				1		5		10		10		5		1				
			1		6		15		20		15		6		1			
		1		7		21		35		35		21		7		1		
	1		8		28		56		70		56		28		8		1	
1		9		36		84		126		126		84		36		9		1

A) Can *you* create a shape of numbers and name it after yourself?

B) How do you feel about yourself now?
1. better
2. worse
3. the same
4. other (Please explain in the space provided.)

#13. Cookies

At the supermarket, Andy buys 144 eggs.

Wait, he thinks. *I'm not invited to Natalya's family party this year. I don't need to make those 18 batches of cookies.*

He'll miss the road trip out there, sipping coffee from the little place they found at the halfway point, listening to Natalya sing the wrong words to his favorite song, "Scar Tissue" by the Red Hot Chili Peppers.

He used to imagine the two of them walking down the aisle. Now the only aisle he walks down is aisle 8 at the market. On the list of things he needs? Eggs, milk, and someone to hold his heart in their hands. He thinks to himself, *What's done is done and I'm working through things the best I can, with the birds I'll share this lonely view.* He's finding new hobbies and engaging in activities that make him feel peaceful, like baking cookies.

A) If it was 144 eggs for 18 batches, how many eggs does he need now that he's baking for just himself?

B) Lost in thought, Andy slips and drops all the eggs on the ground. They cost $3.19 per dozen. How much does he owe?

TO SOLVE — Cookies

A) Since Andy is alone, he only needs to make 1 batch. It's fitting: 1 is the loneliest number, but it also happens to be the easiest number to use mathematically.

Set up a proportion:

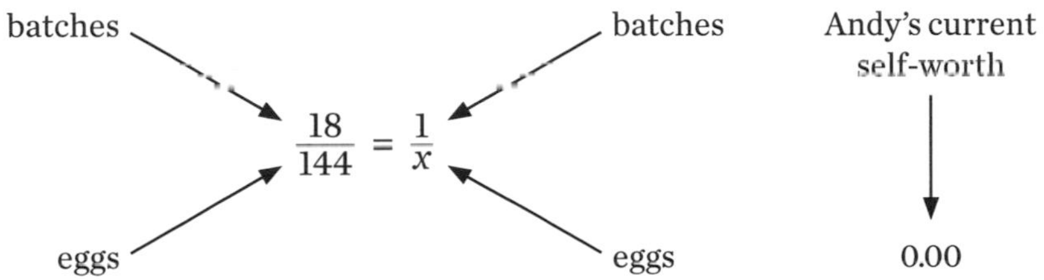

Now you'll want to cross-multiply that.

Cross-multiply on the diagonal.

Set those results equal to each other.

Here's what it looks like:

$$18x = 144$$

You want to make sure x is all by itself (like Andy).

So you divide 144 by 18 (the age they were when they met).

Yep, $x = 8$ (the number of eggs he needs to make cookies for himself).

B) 144 eggs ÷ 12 eggs in a dozen = 12 dozen.

$3.19 * 12 = $38.28. He owes $38.28, plus tax.

#14. Happy Pond, USA

The number of cases of chronic respiratory disease in Happy Pond, USA, is directly proportional to the number of days that the Kurt Midas Oil Corporation spends fracking in the Grandma River. There were 50 cases of respiratory disease this year and Kurt Midas's lackeys were out there fracking for 14 days in June.

But 2 weeks wasn't enough of that sweet, sweet oil money for Kurt Midas. He's got plans to build an infinity pool between the ninth and tenth holes of his personal golf course in Sun Valley.

How many cases of chronic respiratory disease can be expected next year if Kurt is able to have his boys out there for 365 days, rain or shine, come hell or high water, like he promised his investors?

CHAPTER SUMMARY

Congratulations! You've survived Chapter One (can't say the same for that Half-Man Half-Shark). Let's take a moment and review some things we've learned:

1. You remember what Math is. Good work.

2. To find the unknown value, *x*, a series of steps is employed to systematically remove *x* from everything it's attached to, so that we can view it in solitude, one of the worst places to be.

3. Math can be applied in situations both emotional and creative, so maybe "being a creative type" isn't why you don't feel confident in Math.

4. Math is difficult, but so is finding the Big Blue Crystal of Atlantis.

5. The answers are in the book! (Some are in the back.)

REVIEW QUESTIONS

1. Does **McNiel** remind you of anyone you know? What would you say to them if they were here right now?

2. What is the solution to $2x - 3 = 7$?

3. What are some of the unknowns in your life?

4. What is the solution to $\frac{x+1}{12} = \frac{5}{6}$?

5. What's your favorite way to isolate yourself?

Chapter Two

GEOMETRY

MEASURE THIS STUFF, THEN PRETEND IT WILL NEVER CHANGE

Welcome to Geometry, the Math you can see. Geometry is a favorite of many students because of its user-friendly images and easy-to-use formulas. If math classes were stores, Geometry would be Ikea.

Where did it come from?

Geometry was developed by the ancient Greeks, the people who brought you the scientific method, storytelling conventions of "comedy" and "tragedy," and modern democracy. What a great group of guys! Hold on, the Greeks also owned slaves, sequestered women in the home, wiped out civilizations, celebrated binge drinking, and worshipped gods who regularly took advantage of mortals . . . so maybe not.

The early Greek mathematicians were committed. They didn't think about much else besides Math. Day after day, they grabbed their rulers, left their wives and children at home, and went off to spend their days looking at shapes.

How does it work?

Derived from *geo*, meaning "earth," and *metron*, meaning "measurement," Geometry is quite literally the study of earth measurement. Its formulas relate one part of a shape to another part of the shape. If the length of a side is *this*, the area must be *that*. If the radius is *this*, then the volume must be *that*.

Unfortunately, these formulas are useless in most situations. They work for the perfect circle, the perfect rectangle, the perfect sphere, etc., but when it comes to an almost-perfect circle, an 89° angle, or a triangle with curved edges, all these formulas will be off. So Geometry might be the most helpful mathematical tool at your disposal, unless you encounter anything less than perfect.

What's in this chapter?

As opposed to the last chapter, which was about finding the unknown, the problems in this chapter are about getting an answer based on something already known, like a radius, side length, or volume. They're perfect for you if you buy into the fantasy that the things you think you know to be true will never be revealed as false. Go ahead: Measure this stuff, then pretend it will never change.

Geometry is here to help you measure the shape of your world. But can it measure the length of time needed to grieve for a loved one? The volume of the black hole inside of us? The area and circumference of our personal space? Probably not, but after all the years you've put in, the least Math could do is try.

#1. Square One

When you failed that test, your math teacher, Mr. Newman, said in his notes that you'd have to go back to square one.

Here it is:

Think of it as a fresh start—maybe now you can finally make some real progress.

But your rival has found you. "I bet you can't calculate the area of square one!" jeers **McNiel**, appearing in the doorway.

"Oh, **McNiel**! How nice of you to come in during your lunch break to help," notes Mr. Newman. "You sure are lucky to have a friend like **McNiel**. Why, just last week, I saw you two racing bikes. What fun. Anyway, the length of each side is what you get if you divide the number of degrees in a right angle by the number of sides in a pentagon."

How big is square one?

TO SOLVE — Square One

To find the area of square one and prove you are not a failure, you must know that a right angle is equal to 90° and a pentagon has 5 sides. Also, you must not correct Mr. Newman, and instead must pretend to be friends with **McNiel**. Don't let anyone know you care. About anything. Plus, you really don't want to watch Mr. Newman try to handle conflict resolution.

$90 \div 5 = 18$, so the length of each side is 18.

Find the area of a rectangle by multiplying its base by its height (see the Reference Section, page 19). In a square, the most special kind of rectangle, the base and height are equal.

$$18 \times 18 = 324$$

or

$$18^2 = 324$$

324 square what?

(Math books don't need to specify context.)

Hey, let's choose our own! Like inches. Or feet. Or "wibbles."

Yeah, 324 square wibbles.

#2. Andy Builds an Invisible Fence

One part of Andy's life is still good: he got to keep custody of Digit, the dog he adopted with Natalya. He wants to install an invisible fence in the yard so he can let Digit play outside without fear of Digit running away.

The fence can surround a rectangular area of 180 square feet. Andy wants the long sides of the rectangle to be 15 feet, because if he was any farther away from Digit, it would just feel like losing Natalya all over again.

How long should Andy make the shorter sides of the rectangle?

#3. Your Chance to Draw (Now That Funding for the Arts Has Been Cut)

A) Draw a TRIANGLE.

B) Draw a HEXAGON.

C) Ready for a challenge? Draw a TRISKAIDECAGON.

D) Finally, draw a CIRCLE.

#4. True and False: Shapes

	T	F
#1. A circle is a shape that measures 180° total.		
#2. A rectangle is a shape with four right angles.		
#3. A "regular" polygon is a shape with congruent sides.		
#4. "The Shape of You" (from the hit album ÷) is a good song.		
#5. A rhombus is a type of square shape.		
#6. Similar parallelograms are the same shape and the same size.		
#7. A four-sided shape is called a quadrilateral.		
#8. The earth is in good shape.		
#9. You are in good shape.		
#10. When you are in good shape, you are better than other people.		
#11. You better shape up, 'cause I need a man, and my heart is set on you. You better shape up, you better understand: To my heart I must be ______.		

#5. Ninth-Price '90s Media

Natalya wants a more fulfilling career; her BA in psychology is not helping her in the job market. She dreams of starting her own nonprofit, but the thought of sitting down to work out the math for such a thing makes her throat close up.

She's a modern woman, holding down multiple jobs: one delivering for Mr. Houdini Prescriptions and one working at Ninth-Price '90s Media, a store that sells material published in the 1990s at one-ninth the original price.

Currently, Natalya is stacking DVDs:

Never Been Kissed ($2.22);

I Know What You Did Last Summer ($2.44);

and *Mary-Kate and Ashley's Passport to Paris* ($0.89).

Currently the volume of her cylindrical tower is 10,000 cubic centimeters.

Each disc has a radius of 6 cm and the stack needs to be exactly 90 cm tall.

Does she have space to add *Mary-Kate and Ashley's Passport to Paris*?

TO SOLVE — Ninth-Price '90s Media

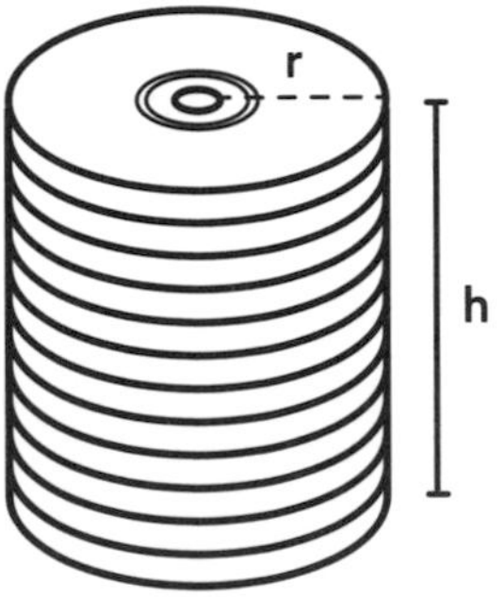

Here's the formula for the volume of a cylinder: $V = \pi r^2 h$.

In other words, the volume is equal to the area of one DVD (πr^2) times the height of the stack.

Now plug the given information into the formula to figure out how tall the stack is so far:

$$V = \pi r^2 h$$
$$10{,}000 = \pi(6)^2 \cdot h$$
$$10{,}000 = 113.097 \cdot h$$
$$88.42 = h$$

Natalya's tower is just over 88 centimeters tall—she'll get in trouble if she doesn't add another 1.58 centimeters. DVDs are only about a millimeter (.1 cm) thick, so yes, she definitely has room for a timeless coming-of-age classic.

#6. NSFW

Sonya and Bart have started a workplace romance, and they've reached a relationship milestone: She found porn on his computer.

The below image represents the relationship between Bart's angle and Sonya's angle when it comes to porn.

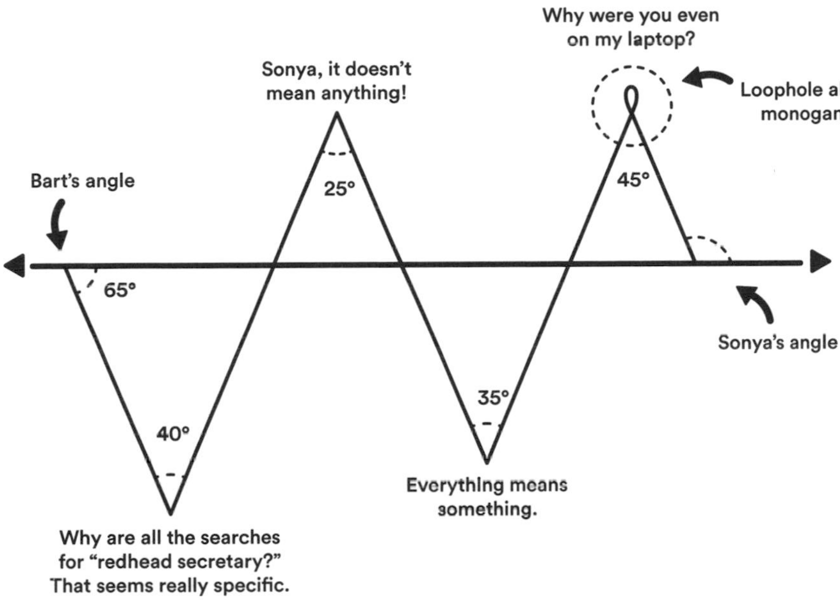

If Bart's angle is equal to 65°, what is Sonya's angle?

TO SOLVE — NSFW

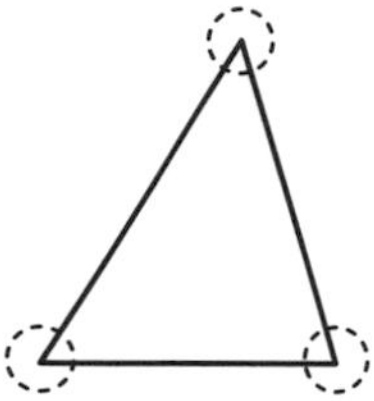

Three angles of a triangle

Angles that combine to make a straight line

Each of the above combinations of angles will always add up to 180°.

Of course, this is true only if you accept that these drawings are perfect. Maybe you should accept them even though they are not perfect. Even if they watch porn, or sometimes borrow your stuff without asking.

Also, Timberlake's angles will always be equal to each other:

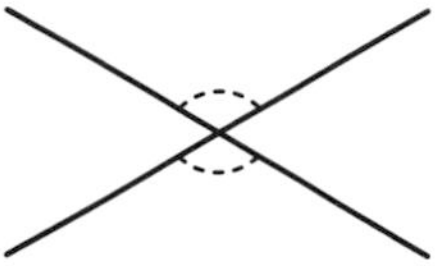

Timberlake's angles

Note: "Timberlake's angles" are often misnamed in math classes as "vertical angles." The correct name is "Timberlake's angles," because it's like they're my mirror, my mirror staring back at me.

Use the triangles, straight lines, and Timberlake's angles in the diagram to fill in the missing angles.

Sonya's angle is that she feels she deserves to be the most important woman in Bart's life, and doesn't want to feel like he's keeping secrets from her, all of which equals 110° on this diagram.

#7. Life Isn't a Highway

Mr. Newman races home on a straight highway at 65 miles per hour. His thoughts, however, are racing in circles at a mile a minute. *Should I have just told my students how old I am? No, no, then they wouldn't respect me. People are only respected when they are distant, like Mom.*

As he looks forward, straight into the sunset, he wonders, *Why can't I stop going in circles?*

The circle his thoughts are running in has a circumference of 20 unpleasant memories.

What is the radius of the circle (measured in unpleasant memories)?

#8. The Box

A rectangular box is 30 inches long, 30 inches wide, and 10 inches high.

How many identical rectangular boxes (each 3 in × 3 in × 1 in) can fit inside?

(Hint: Be careful—don't think outside the box.)

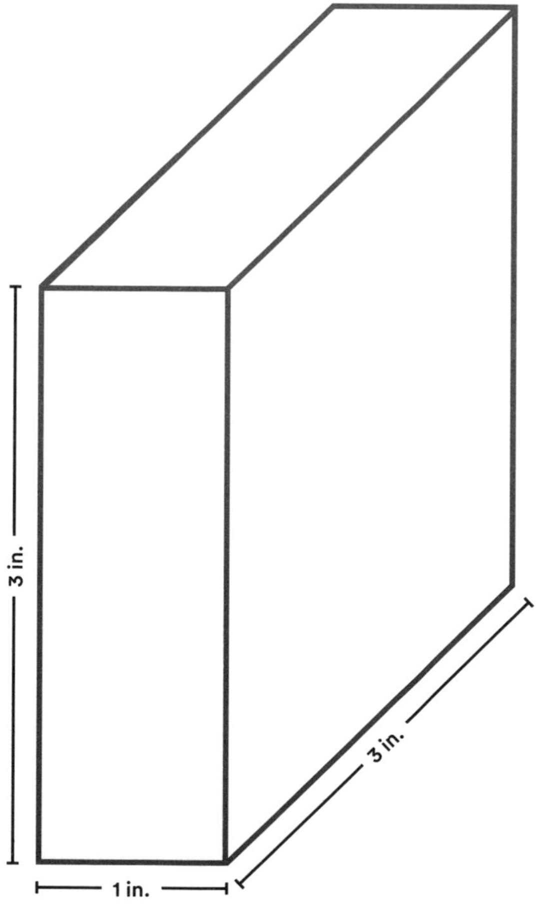

#9. The Rock Show

You need to cover a rock with fabric, because this is a math problem. The rock is spherical in shape, with a diameter of 2 inches and a blissful ignorance of your feelings.

A) How much fabric do you need?

After solving that, you become very frustrated. What shape do you cut the fabric in? How do you glue it together? Do you need to sew it? You don't know how to sew! Eventually you decide to just bike to **McNiel**'s house, throw the rock through your rival's window, and ride away. Not your finest hour, but hey, it's easier than doing Math.

B) What is the area of the hole the rock left in **McNiel**'s window?

(Note: Assume that the rock leaves a perfectly circular hole, not because that can realistically happen but because, again, this is a math problem.)

FACTS & OPINIONS (FROM MIKE)
π

The number π is one of the most remarkable numbers in all of Mathematics.

Specifically, π is the ratio of a circle's circumference to its diameter. If a circle's diameter is 2 feet, then its circumference is 2 π (roughly 6.28) feet.

π is considered to be an "irrational number," which means that its decimals go on forever, with no repeating pattern.

Archimedes, a Greek mathematician, was the first to accurately calculate π—he did so by using geometric techniques.

In 1761, Johann Heinrich Lambert proved π to be irrational. His proof was a beautiful representation of the eternal perfection of Math, as it represented a coming together of many branches of Mathematics. The proof, while on the subject of π (which is most useful in Geometry), was actually based on Trigonometry and Calculus, which just goes to show how linked all these concepts really are, no matter how disconnected they may seem at first.

π remains prominent in contemporary math subculture. Contests are held to see who can memorize the most digits of π. The strongest supercomputers' algorithms are often tested by calculating as many digits of π as possible. We have now calculated more than a trillion digits of π, and there is still no end in sight.

OPINIONS & FACTS (FROM TAYLOR MARIE)
π

The number π is one of the most ludicrous numbers in all of Mathematics.

Specifically, π is only used for circles, is actually a *letter*, doesn't make any sense, and gets its own holiday.

You see, when things are irrational we give them a holiday, like Jesus rising from the dead, a little pot of oil burning for over a week, people dying in wars, and π.

Archimedes, some Greek megalomaniac, brought us this fun concept—he did so by using his time calculating numbers instead of connecting with other people.

Let's imagine the guy who figured out that π went on forever, never repeating itself. It must have been one of the greatest existential crises ever to have happened. How many digits did he get through before he started to accept they were never going to repeat? A metaphor, perhaps—for the value of living in the moment? That no one moment shall ever repeat again? That night, perhaps he looked up at the sky or out at the sea or whatever, and thought, *Oh god of my choosing, I pray that one day people will celebrate this moment by eating dessert.*

π remains prominent, and kids from across the globe use it as an opportunity to prove they are better than other kids. We use π to test supercomputers, just like we use it to test children in school, because humans and machines should be treated equally.

#10. The Offer

Anita the farmer just got the following letter:

To Whom It May Concern:

Congratulations. I, Kurt Midas, president and CEO of the Kurt Midas Oil Corporation, am interested in purchasing your farm. Sources tell me there is oil under your land, and I'd like to own that oil and then sell it to others. I'm prepared to make you the generous offer of 10 cents per square foot of land. You'd be wise to accept, as I might not feel so generous this time next year.

I look forward to doing business to you,

Kurt Midas
President and CEO, Kurt Midas Oil Corporation

The last thing Anita wants to do is sell her land, a 3000 × 4000-foot rectangle that has been in her family for generations. She sighs, massaging her temples with rough fingertips. Every day it gets harder to support herself and her daughters with the proceeds from her farm.

What is the monetary value of Kurt Midas's offer?

#11. New Apartment

Now that Andy isn't paying half her rent, Natalya must find a smaller apartment. Who knew personal space was so expensive?

Natalya is looking forward to something that is just hers. She can use the space to start her own business: a nonprofit that turns cigarette butts into school supplies for children.

The first image represents the floor plan Natalya saw in her mind. The second image represents the floor plan she actually saw today.

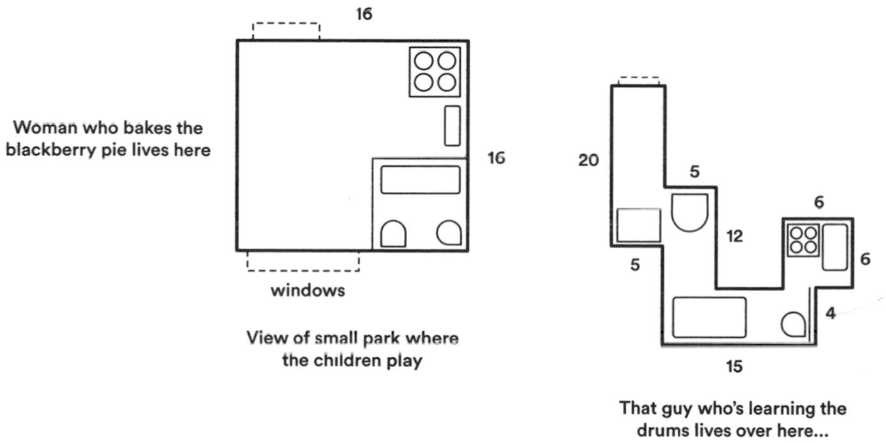

Do these two apartments have the same square footage?

TO SOLVE — New Apartment

The first one is pretty straightforward: 16 ft by 16 ft is 256 square feet.

For the second one, divide it into rectangles:

The four rectangles have areas of 100, 60, 60, and 36 square feet.

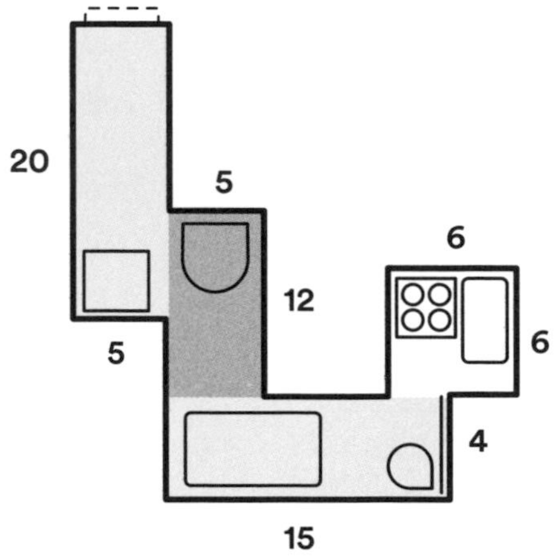

Add those together and we have . . . 256 square feet.

These two apartments are exactly the same.

#12. The Inner Circle

Norah's all alone in her big big house. It's just her, her thoughts, and a joint of Kevin Wake'n'Bacon. Her husband is away on business, and her son is at Virtual Reality Camp. She falls back on her bed, lets her mind unwind, and gazes through her circular skylight.

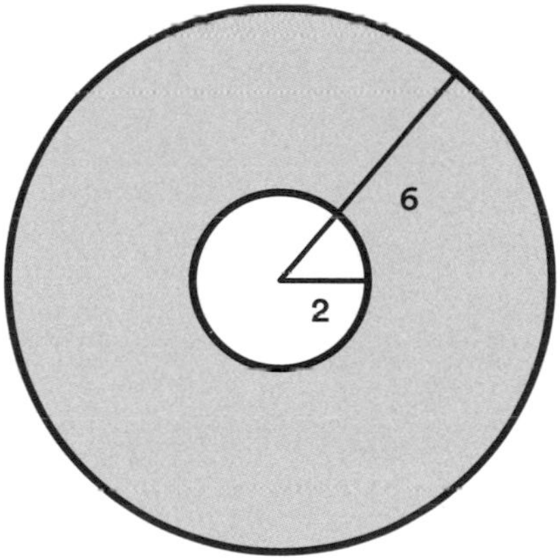

She notices the outer circle is bigger than the inner circle. Ain't that the truth. Norah spent years on the outer circle, like Kevin Bacon at the beginning of *Footloose*. But all she wanted was to be in the inner circle, like Kevin Bacon at the end of *Footloose*. Now she's finally made it: a big big house, cool friends, the works.

Still, the inner circle feels so small, so cold, and so sad, like Kevin Bacon in *Mystic River*.

If the radius of the inner circle is 2 wibbles and the radius of the outer circle is 6 wibbles, what is the area of the shaded region?

#13. Dog-Gone It

Digit the dog was just hit by a car going 65 miles per hour. Even though Andy correctly calculated the size of the invisible fence, he made a mistake setting up the electrical wiring, since he had received no education in that area.

Now Digit,

his best friend,

lies mangled in the street.

Andy enters the first stage of grief: logistics. He must buy a coffin and both literally and metaphorically bury his relationship with Natalya deep within the ground. He walks into the store and sees he has two options: a big coffin (shown below) listed at 8 cents per cubic inch, or a big coffin of the same size listed at $999.99.

Which dog coffin is the better deal?

Hint: $V_{coffin} = \frac{(b_1 + b_2)(l)(h)}{2}$

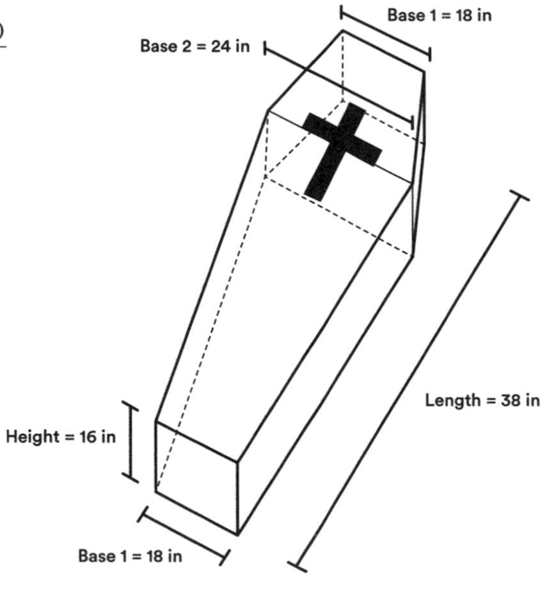

#14. Quest for Salvation

Cody is really enjoying Virtual Reality Camp. At first he was afraid he might miss his mom, but at 12 years old he's becoming a man, and men do not miss.

In his current simulation, Cody stands atop a hill, looking down on chaos: fire, screaming, the pungent scent of gunpowder. The setting sun beams on a battlefield soaked with the blood of humans and aliens alike.

"General!" cries a boy carrying his own severed arm in his backpack as he rushes up the slope. "There are too many of them! We need to escape; our only chance is to cross the water in those boats!"

The following map appears on Cody's user-interface:

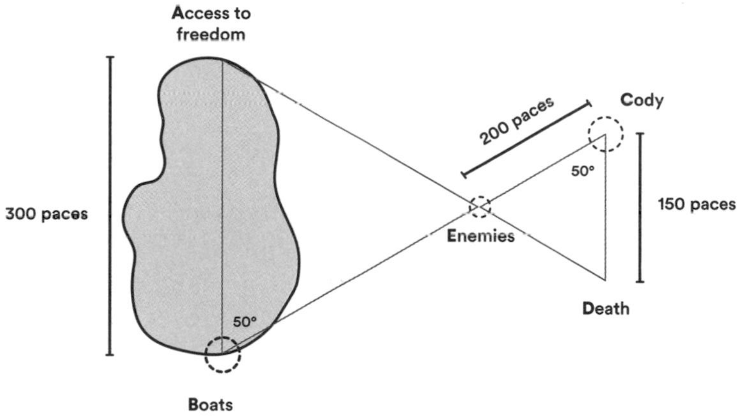

How far is Cody from the boats?

TO SOLVE — Quest for Salvation

There are two sets of matching angles in the triangle:

1. Angles B and C (both 50°)
2. The two angles at point E (Timberlake's angles)

Since triangles always add up to 180°, angles A and D must be equal as well. Does this mean death and access to freedom are equal? According to this math, yes.

Triangles that are the same shape but different sizes are called similar triangles.

Their sides are proportional to one another, and they have the same hobbies.

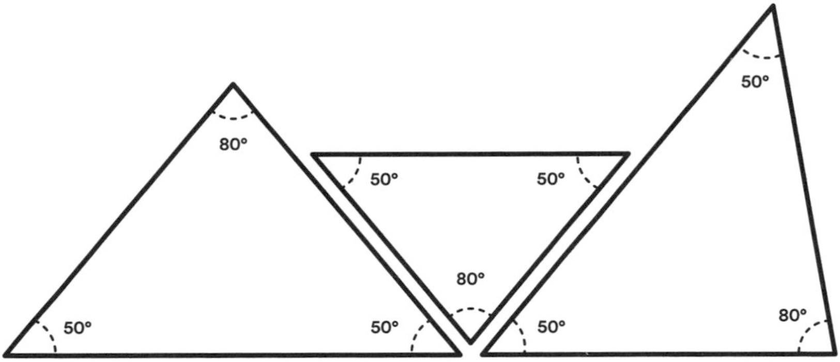

So since AB is twice as long as CD, that means every side in triangle ABE will be twice as long as every side in triangle DCE.

CE is 200 paces and corresponds to BE, which must be 400 paces.

So Cody needs to run 600 paces to get to the boats. Or rather, Cody needs to not move at all, but hold down X on his controller for a little while.

CHAPTER SUMMARY

Geometry is over, and with it the life of Digit the dog. This chapter is written in your memory, Digit. You were not real, but you were loved. Let's go over what we learned:

1. The volume of a cylinder is equal to the area of the base times the height.

2. Geometry was invented by the Greeks, who may not have been good people.

3. The economy is a major determining factor in the size and shape of someone's personal space.

4. π is the ratio of a circle's circumference to its diameter.

5. Some things can't be measured, like grief, love, and the hole inside the heart of someone who has experienced both.

REVIEW QUESTIONS

1. Which is larger: a square with 6-foot sides, or a circle with a 6-foot radius?

2. What is the radius of the moon?

3. Measure the square footage of your bedroom. Now measure the square footage of your roommate's bedroom. Are you paying the right amount for rent?

4. What role would you rather see Kevin Bacon play: an aging rock star, a hard-nosed football coach, or a closeted gay surgeon who moves back to his hometown in Alabama only to find that his favorite bakery is getting shut down, so he assembles a ragtag group of his high school buddies to put on a musical to raise money to save the bakery?

5. Memorize 3 digits of π and not one more. Unless you are a rocket scientist, in which case please memorize 8.

Chapter Three

TRIGONOMETRY

WHAT'S YOUR ANGLE?

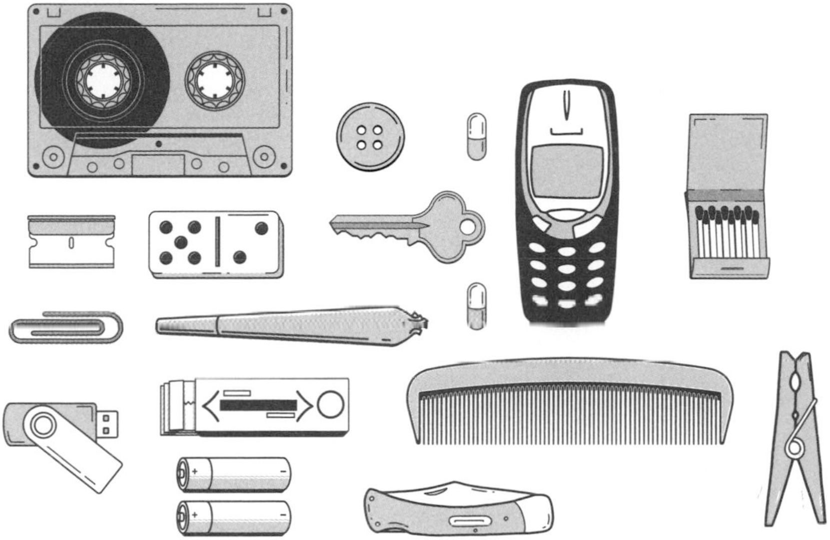

Trigonometry is the study of triangles, the spookiest shape. No shape, except maybe *The Ring*, has ever been so spooky. The Illuminati, the Bermuda Triangle, the three branches of government . . . all very spooky.

Trigonometry shows up in students' lives when they least expect it. Trig doesn't fill up a year-long class on its own, so it pops up, uninvited, in all sorts of places. You might remember Trigonometry from your Geometry class. You might recall it from your Algebra 1 class, or you might have seen it in Algebra 2, the poorly reviewed but high-earning sequel to Algebra 1 that nobody asked for.

Trigonometry is how we know how far away from the moon we are, and also how we create wavelengths for satellite radio. So it's pretty much useless, unless you like living in a house, having a cell phone, getting the occasional MRI, or using bridges.

Where did it come from?

Trigonometry emerged in Hellenistic Greece during the third century BCE. Hipparchus is credited as the father of Trigonometry, probably because he spent the most time looking at the sun and moon, thinking, *How far away are those things? Why does it feel like they know everything about me and I know nothing about them?*

How does it work?

Trig is largely recognized by its mysterious mnemonic device: **SOH, CAH, TOA**

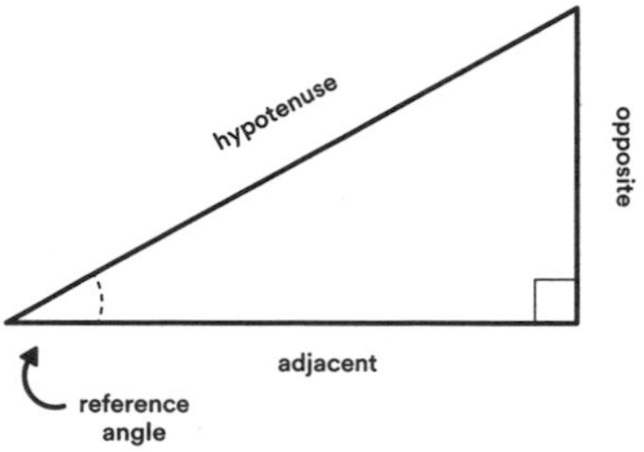

$$\text{SOH: sine} = \frac{\text{opposite}}{\text{hypotenuse}}$$

$$\text{CAH: cosine} = \frac{\text{adjacent}}{\text{hypotenuse}}$$

$$\text{TOA: tangent} = \frac{\text{opposite}}{\text{adjacent}}$$

Trigonometry uses known ratios of sides in a triangle called sine, cosine, and tangent to measure unknown lengths, heights, and distances. You may be staring longingly into the void, wondering, *How can I know my opposite if I don't know myself?! How can I know what's adjacent to me if I don't know where I am?* In math education, this is known as finding the "reference angle." In therapy, it's known as "self-awareness."

For a given angle, the three ratios (sin, cos, tan) remain constant no matter what. The sine of 30°, for instance, is equal to $\frac{1}{2}$ in a triangle of any size, from the microscopic to the galactic. Therefore, you can use Trigonometry to measure how tall your desk is, how tall your mom is, or how tall a mountain is.

You see, dear friend, Trig offers you a powerful gift: the chance to gain information about things without directly engaging with them. The people who determine which ads you see on social media know lots of Trigonometry.

What's in this chapter?

This chapter is full of triangles of all kinds. The questions will be all about those. Is this triangle acute? Is it obtuse? What is the sine? What is the cosine? Do you need someone to cosign your lease? Is the rent too damn high? Do you need to get a second job? When would you have time for that? If the federal government shuts down for a week, should that affect your taxes? Are we going off on a tangent? Apparently not.

In this chapter, goals are on the horizon, but dreams feel far out of reach. Use Trigonometry to figure out *exactly* how far, and everyone will feel better, right?

#1. Find Sine, Cosine, and Tangent

You're at a dinner. The person to your left has been talking for 10 minutes about how much better their life is since getting rid of their TV, and the person on your right is engaged in taking "portraits" of the salmon.

You'd rather not sit here and try to understand these people's angles. You excuse yourself, go to another room, and take out your math homework, which, unfortunately, is full of angles.

Solve the following:

A) $\sin(60°) = ?$

B) $\cos(45°) = ?$

C) $\tan(20°) = ?$

TO SOLVE — Find Sine, Cosine, and Tangent

If you know your reference angle, you can use your calculator to find the sine, cosine, or tangent.

A) Let's find the sine of 60°.

1. If you're using a smartphone calculator, turn it to the side so the extra buttons appear.

2. Type in [6] [0].

3. Hit the [sin] button.

Did you get .866? Don't panic—we know decimal places can be scary since they are a reminder that things are not perfect. You really only need to use 3 decimal places, even when they go on and on and on. Seriously, for your sanity, just look at the first 3 decimal places, and try not to start thinking about "forever." It will certainly get in the way of doing the math. .866 is correct.

If you got −.3048, it means your calculator is in radian mode. Don't be ashamed; hit the button marked deg to switch into degree mode. (Radians are just another way to measure angles, sort of like how feet and meters both measure distance.)

Try the other buttons to see:

B) cos(45°) = .707

C) tan(20°) = .364

Don't forget to thank your pocket robot for all its hard work.

#2. Funding

Natalya takes a deep breath and looks up at the top floor of the Ball, Shaft, & Associates building, where she has a meeting today with potential funders for her organization. Her idea, a nonprofit that turns cigarette butts into school supplies for children, seemed great, but now that she's only 40 feet from the door, she's sweating through her blazer.

She wonders, *How many entrepreneurs are the Associates meeting with this quarter? Has someone else already come up with my idea?* There's no way to know. All she can do right now is hold her binder tight and look up at an angle of 81°.

Literally how far out of her grasp is the dream of getting funding for her company?

TO SOLVE — Funding

Let's visualize Natalya's situation:

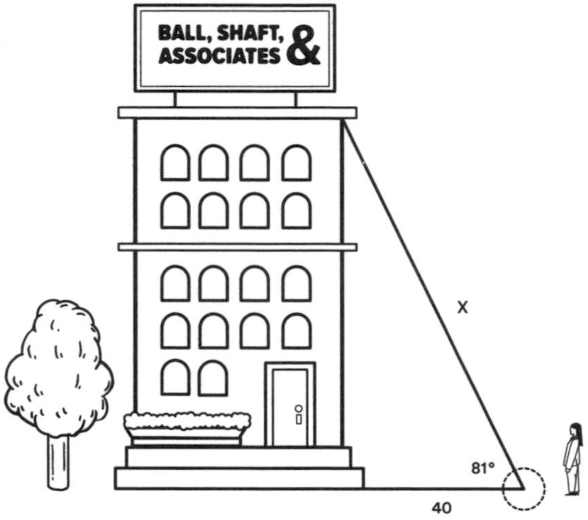

The adjacent is known, and the hypotenuse is unknown. (The opposite is also unknown, but it is not affecting Natalya's sense of worth, so we don't need it.)

Therefore, we use cosine = $\frac{\text{adjacent}}{\text{hypotenuse}}$

$\cos(81°) = \frac{40}{x}$

$.156 = \frac{40}{x}$

$.156x = 40$

$x \approx 256$ ft

Her dream is 256 feet out of her reach.

For now.

#3. Periods

The gray line is a graph of regular periods, described by $y = \cos(2\pi x)$.

The blue line is a graph of irregular periods, described by a person.

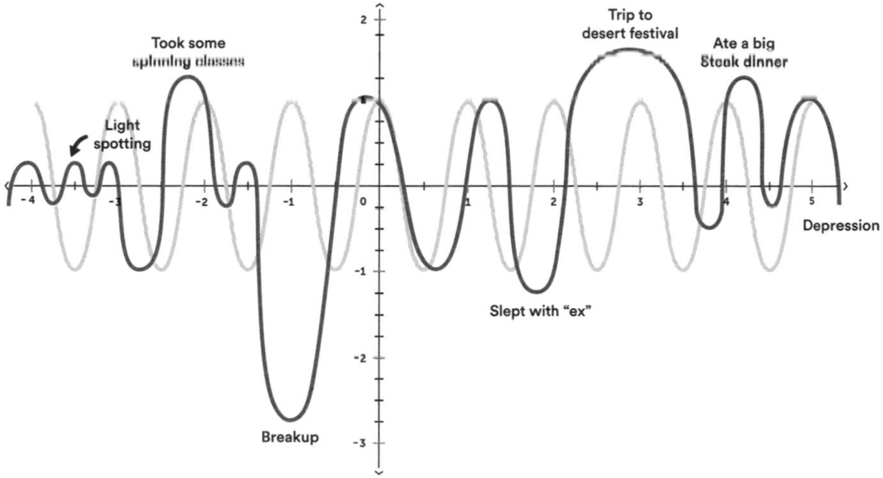

If the amplitude of this person's period is not normal, the blue line will have a higher peak than the gray line, or a lower valley than the gray line.

When her period is not normal, she'll take a pregnancy test.

Over the 9 months shown on the graph, how many pregnancy tests will she take?

#4. The Perfect Angle

Andy is trying to take the perfect selfie for his profile on Crumble, a dating app. He wants to make sure it looks good, so he'll only use a photo taken from an acute angle. He takes selfies from the following angles:

60°
90°
120°
150°

Which one can he use?

#5. See Your Rival in a New Light

Your rival, **McNiel**, stands alone in the moonlight, holding the rock you threw through the window and crying, silently, into the dark.

A single moonbeam shines on the rock,
    Illuminating it.
Perhaps it has some kind of inner life?
    Of course not.
Like the moon itself, it is a rock.
    It knows nothing of "inner life."

The light catches the attention of a nearby nighthawk, who swoops down, snatches the rock from **McNiel**, and flies away at a 20° angle from the horizon. It perches atop a tree, the base of which is 200 feet away.

How tall is the tree?

#6. Quarterback in the Day

Still alone in the big big house, Norah's in the big big bubble bath with a joint of Sativa Gomez. Her husband hasn't called from his business trip, and her son hasn't written her a letter from Virtual Reality Camp. She thinks back to when she was at camp playing football.

10 seconds left on the clock. Ball on the opponent's 30-yard line. Time for one more play. Young Norah takes the snap, drops back 5 yards, and rolls to the right. As the buzzer sounds, she turns 50 degrees and throws a bullet pass to the wide receiver, who had run a fly pattern down the left sideline. The receiver catches young Norah's perfect spiral on the goal line, and her team wins.

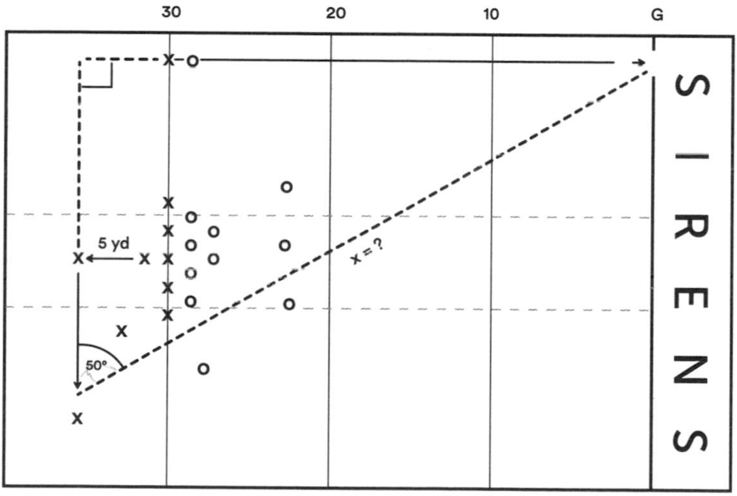

A) How far (x) did Norah throw the ball?

B) If football is the most popular sport in America, why aren't women allowed to play it professionally?

FACTS & OPINIONS (FROM MIKE)
ERATOSTHENES

Eratosthenes was a well-known Greek mathematician/geographer/ poet/astronomer/music theorist.

Pretty impressive résumé:

- Chief librarian at Alexandria
- Established a simple algorithm to find prime numbers
- Created latitude and longitude
- Invented Leap Day

Eratosthenes's most notable achievement might be his calculation of the circumference of Earth.

Eratosthenes knew of a well in Syene, Egypt, where at noon the sun was directly overhead. One day, at noon in Alexandria, he used a small rod (called a *gnomon*) to measure the sun's angle of elevation there. Seeing that the angle was about 7.2°, which is $\frac{1}{50}$ of a circle, he concluded that Earth's circumference would be about fifty times the distance between Syene and Alexandria. He knew this distance to be about 5,000 stadia, so he estimated the circumference of Earth to be 250,000 stadia.

Because it is not clear exactly what Eratosthenes meant by "stadia," history is unclear as to the accuracy of the estimate. The optimistic interpretation shows that he was *less than one percent off* from the actual circumference of Earth. The pessimistic one shows that he was about sixteen percent off. Still—not too shabby for 240 BCE.

OPINIONS & FACTS (FROM TAYLOR MARIE)
ERATOSTHENES

Eratosthenes is one of those old Greek guys with a long beard and a long name that sounds like a prescription drug.

Pretty big know-it-all:

- Chief librarian at Alexandria
- Pretentious poet
- Felt the need to be able to locate everything on Earth
- Felt the need to invent a new day

Eratosthenes's most notable achievement might have happened by accident.

Eratosthenes had a favorite thinking spot in Syene, Egypt: a well he liked to cry into when he became overwhelmed by the burden of who he was. After being rejected by a woman speaking her truth, he went there with a small rod to thrash angrily at the bushes, shouting in Greek, "She doesn't understand me, so that woman will never be happy!" He threw the rod, which stuck into the ground. The sun was not shining on him metaphorically, but literally it was, and it gave the rod a shadow with an angle of 7.2° ($\frac{1}{50}$ of a circle). Distracting himself from his heartbreak with logic, he concluded that Earth's circumference must therefore be fifty times the distance between Alexandria and Syene.

Today we can prove that his estimation was between one percent and sixteen percent off target. As for what he said about the woman . . . no one asked her how she felt, 'cause no one in ancient Greece cared.

#7. Falling Apart

Yesterday, Andy took a leap: He met someone on Crumble and asked her out on a date. Now he's about to take another leap: out of a plane, because they're skydiving.

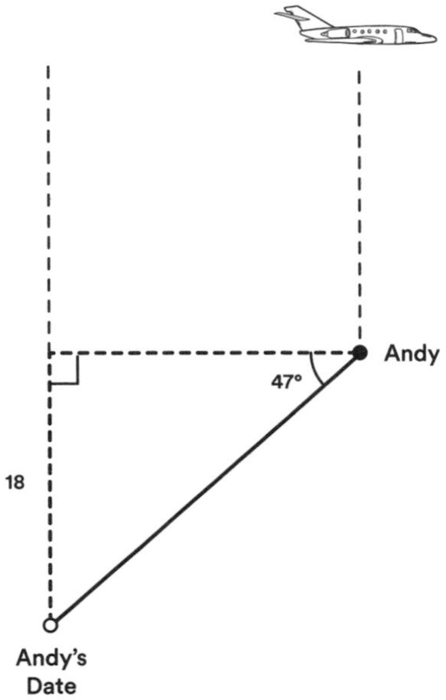

He thought they'd share a parachute, but it turns out that's illegal. Now his date is harnessed to a big, strong instructor with a great sense of humor, and boy are they hitting it off. Andy is confused . . . they were an 82% match on Crumble.

Laughing loudly at a particularly good joke Andy didn't hear, his date and her instructor bravely jump out of the plane, spooning each other for safety.

The above image shows them falling from the sky. According to these measurements, how far apart are Andy and his date?

#8. Off the Radar

In this simulation, the great ship *Hipparchus* is sailing through murky waters, fleeing from teething baby sea monsters.

Computer-generated wind blowing through his hair, Cody can hear an enchanting song coming from a small island at a 45° angle to his left. He looks through the ship's radar, seen here:

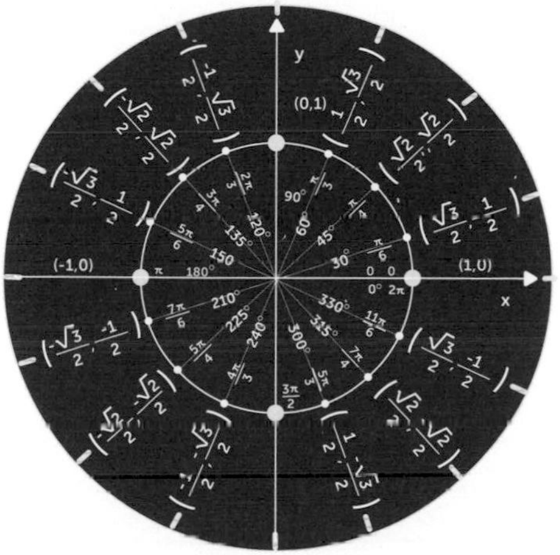

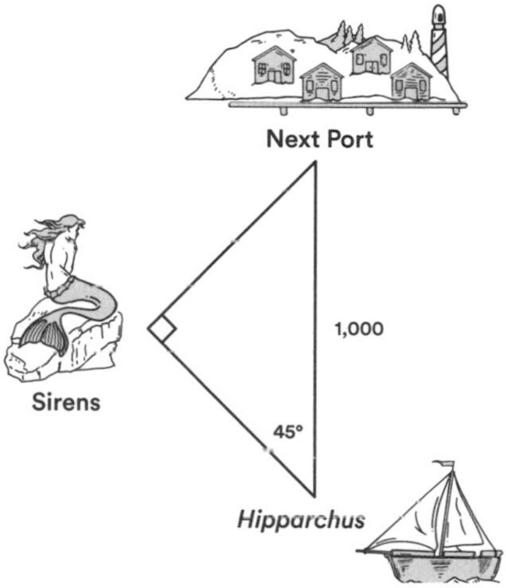

The simulation gives him two options: Stop and engage with the beautiful sirens, or keep his honor and continue piloting his men safely to the next port, 1000 feet away.

A) How much extra distance is it to make just a quick stop to visit the sirens?

B) Wait, why would it be dishonorable for Cody to interact with them?

#9. Bermuda Triangle

An alien needs to abduct a human as an anniversary gift for its partner. It sets out toward Florida from Puerto Rico, searching for someone who looks like they wouldn't be missed.

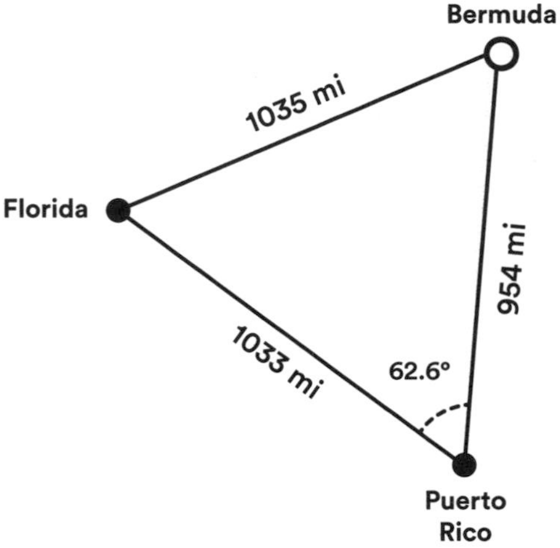

Luckily for this alien, Mr. Newman is spending his weekend at the beach. It traps your unsuspecting teacher in a beam of light, pulls him into the spaceship, turns, and flies in a straight line from Florida to Bermuda.

How many degrees did the alien ship turn?

> Hint: Law of Sines
>
> $$\frac{\sin A}{a} = \frac{\sin B}{b}$$

TO SOLVE — Bermuda Triangle

Use a proportion! Just make sure that angles and sides match accordingly. For instance, the Florida angle corresponds to the 954 mile side. Your proportion might look like this:

$$\frac{\sin(62.6)}{1035} = \frac{\sin B}{954}$$

$$\frac{.888}{1035} = \frac{\sin B}{954}$$

$$(954)(.888) = (1035)\sin B$$

$$\frac{(954)(.888)}{1035} = \sin B$$

$$.819 = \sin B$$

Now, type .819 into your calculator. If you are on your phone, turn your calculator to the side and hit the button marked [2nd].

[sin] will turn into [$\sin^{-1}$]. Hit that button now.

$$B \approx 55°$$

Of course, the alien did this math in its head, or whatever it has where we would expect a head to be.

#10. What's Your Sine?

Aries (Mar 21 – Apr 20)

An exciting and romantic individual will enter your life, ask you for directions to the bank, and exit your life forever.

Taurus (Apr 21 – May 21)

The stars say your insight will be needed in a matter of choosing which movie to watch.

Gemini (May 22 – Jun 21)

Venus and Mars move into your 3rd house, as you move back into your mom's house.

Cancer (Jun 22 – Jul 22)

You will see a meme and think, *Yup, I can relate.*

Leo (Jul 23 – Aug 23)

Your career will take an unexpected turn when Earth turns away from the sun tonight.

Virgo (Aug 24 – Sep 22)

Your patience will be tested today when some dingus blocks the intersection as you're trying to turn left.

Libra (Sep 23 – Oct 22)

With Mercury in retrograde, you'll enjoy listening to music, but not while you exercise. What's that about?

Scorpio (Oct 23 – Nov 21)

You will receive some upsetting news today when you read any news source available.

Sagittarius (Nov 22 – Dec 21)

Now is a good time to reflect on your past mistakes, then keep acting the same way you always have.

Capricorn (Dec 22 – Jan 20)

When looking at the moon this evening, you will feel a tinge of regret for not learning to play the violin when you had the chance.

Aquarius (Jan 21 – Feb 19)

Your solar chart suggests that your DNA patterns are made of 4 basic proteins and that you're a good houseguest, always bringing something for the host.

Pisces (Feb 20 – Mar 20)

Today you will read a great piece of fiction, so clever, so moving, it makes your whole day good.

#11. Not That Kind of Love Triangle

Sonya is having a walk'n'talk with Natalya, venting that she isn't feeling appreciated by her partner, Bart. He's way too focused on making partner at Ball, Shaft, & Associates.

Natalya, who just came out of a very disappointing meeting, is happy to take this rare opportunity to show off her degree in psychology. She describes to Sonya the concept of Sternberg's Triangular Theory of Love, which says that commitment, passion, and intimacy are equally important parts of loving interpersonal relationships. Seen here:

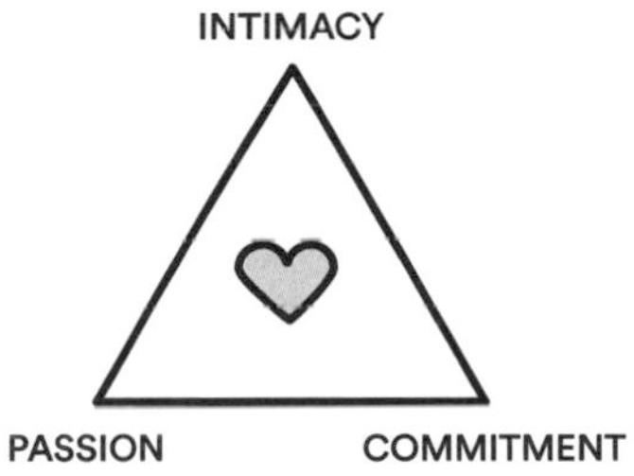

"If the relationship lacks intimacy and passion, it wouldn't be wise to stay in it for the sake of commitment," Natalya explains, mentally reassuring herself that she is indeed happier alone.

Sonya is about to use the Law of Sines to figure out the angle of her commitment to Bart, when she steps on a pencil. Just like the pencil Bart lent her on their first day of work. Suddenly, Sonya throws all logic to the wind and instead applies the Law of Signs.

"Who would leave a pencil just *lying* on the sidewalk?!" she exclaims. "We're meant to be together. Love can't be measured."

"It can, using this triangle," Natalya insists.

Who's right?

CHAPTER SUMMARY

Congratulations! You've completed Chapter Three. We don't have a letter of the alphabet to let you know how you did, but here's a letter from a proud mom:

Honey,
I'm so proud of you.
And I'd like to personally thank you
for never making me have to say
those dreaded words,
"My kid
is struggling in Math."

Let's see what we learned:

1. Sine, cosine, and tangent are all terms for constant ratios that are used to discover unknown measurements in a triangle.

2. Trigonometry is responsible for many things we take for granted, like bridges, geography, cell phones, and love.

3. Skydiving is a risky choice for a first date.

4. In Math, a period is defined as the time it takes for a sine or cosine wave to repeat. In a woman's body, a period is defined as a monthly shedding of the uteral lining with the expulsion of blood and tissue through a small opening in the cervix and into the vagina. So just let her do what she wants, OK?

5. Everything you've ever heard about the Bermuda Triangle is one hundred percent true.

REVIEW QUESTIONS

1. What is the sine of 90°?

2. Look up your astrological sign. Look up your crush's astrological sign. Do some online research to see if you are compatible. Write down how much time you spent on that instead of calling them.

3. List 3 things that Eratosthenes accomplished.

4. List 3 things that Eratosthenes didn't accomplish.

5. Which of the following are obtuse angles?

   A. 30°
   B. 130°
   C. "I'm so sorry, dude, I honestly had no idea it would bother you that I slept with your sister. If I had known, I never would have done it. That's my bad, dude. That is my bad."
   D. 45°

Chapter Four

CALCULUS

THE HELLBEAST'S SURPRISE

You may remember Calculus from high school, your nightmares, or one of its many roles as a dramatic evil in films.

Whatever your experience with Calculus is, there is only one thing you must know: *You don't need it, and you never will.* Like all tyrants, Calculus uses scare tactics to hide the fact that behind all the smoke and mirrors, it is completely unnecessary.

Calculus is *useful*, but only for things that are superfluous and/or annoying. Think of it as the midlife crisis of Math. It is painstakingly complex, impossible to understand without first enduring years of struggle, and only good for expensive and esoteric purposes: rocket science, warfare, working in finance, and spending lots of time thinking about infinity.

Where did it come from?

From the depths of the darkest abyss, born of fire and blood, Calculus rose from the black hellmouth as a savage beast hungry to feast upon the tears of the young. Upon reaching the surface, it set its sights on two bored and highly successful white guys: Gottfried Leibniz and Sir Isaac Newton.

In the late 1600s, Newton and Leibniz, unsatisfied with their mastery of two millennia of Mathematics, set out for immortality. In the 1680s, they both published papers on what is now known as Calculus—pretty much the mathematical equivalent of *Armageddon* and *Deep Impact* both coming out in the summer of 1998. People enjoyed both, didn't really mind that they were similar, and everything was cool. But in 1699, things turned ugly. Nicolas Fatio de Duillier, another bored and highly successful white guy, accused Leibniz of plagiarizing Newton's ideas. They fought and argued about who would be immortalized as the man who invented Calculus until Leibniz died in 1716 and was buried in an unmarked grave.

The hellbeast had claimed its first victim.

How does it work?

Oh, boy. We're so sorry you want to know, but to be polite we'll answer your question. For whatever reason you wish to do calculus, whether it be for your engineering major, to impress somebody,* or to just torture yourself with something complicated, you should know three words: function, limit, and derivative.

* If you are looking to impress somebody, consider doing something else, like learning the bass or shooting a 3-pointer.

1. Function: A relationship between two variables.
Such as: $y = -4.9x^2 + 10x + 20$, which can also be written as $f(x) = -4.9x^2 + 10x + 20$

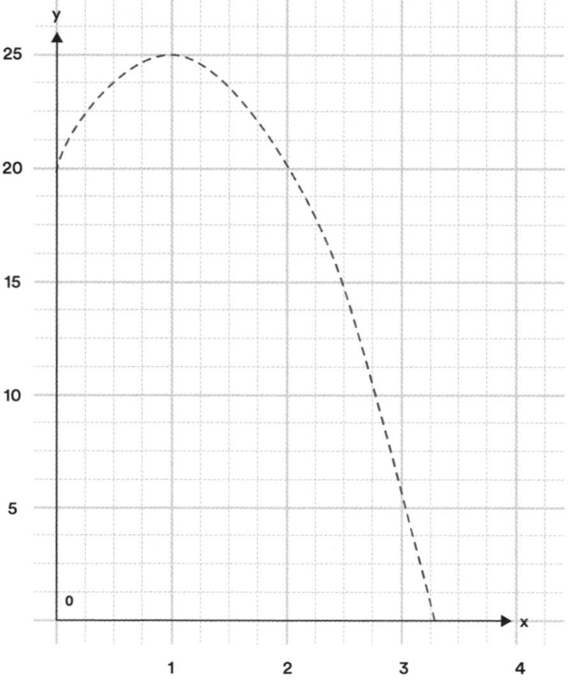

$f(x)$ reads as "function of x," which is the same as y. Here's something else: why does $f(x)$ have so many different channels? There's FX, FXX, and FXM, and aren't Fox News and 20th Century Fox part of that whole thing, too? It's pretty hard to track their moral and political agendas.

The function $f(x) = -4.9x^2 + 10x + 20$ represents you, standing on the edge of your 20-meter-high roof, throwing your turkey sandwich into the air at 10 meters per second (m/sec). (If you're a vegan, it is your tofurkey sandwich with veganaise on sprouted gluten-free toast or whatever. Good for you, by the way.)

20 represents the initial height of the sandwich.
10 represents the initial speed of the sandwich.
4.9 is the effect of gravity (Thanks, Newton!).

On the graph, time is on the x-axis, space is on the y-axis, and you are having lunch on the roof. Looking at the graph of the function, we can see a few things:

- At 0 seconds, you throw the sandwich up in the air from a height of 20 meters.
- At about 1 second, the sandwich reaches its highest point.
- At about 2 seconds, it falls past your face. You've never seen lettuce this beautiful before, waving in the wind, hovering between a tomato and a pair of pickle slices. This must be why in church they say "lettuce pray."
- A little after 3 seconds, the sandwich hits the ground, where it will eventually be eaten by birds or rabbits.

2. Limit: The y-value that a function approaches as it approaches an x-value.

For example: $\lim_{x \to \infty} -4.9x^2 + 10x + 20$

(As a sentence, this reads: "The limit, as x approaches infinity, of the function $y = -4.9x^2 + 10x + 20$."). It means: When x gets close to infinity, where is y?

Let's test it by making a little chart, and see what happens to the y-value as the x-value increases. Did you get scared when you read the word *test*? Just put a pin in that. Using a calculator, we plugged these values of x into that long number sentence about you and your sandwich.

x	$y = -4.9x^2 + 10x + 20$
1	25.1
10	−370
100	−47,980
1,000	−4,889,980
1,000,000	−4,899,989,999,980

So here, x is approaching infinity and the y-values are getting very negative indeed, like that one friend who's always saying stuff like "It's a shame there's no place for art in modern capitalism." Here the y-value is approaching negative infinity. Therefore, we can say: $\lim_{x \to \infty} -4.9x^2 + 10x + 20 = -\infty$

Everything seems to be approaching infinity. You, your turkey sandwich, and the universe.

We can also see the limit on the graph of this function. See how as the line goes to the right on the x-axis (toward infinity) it also goes down at a steeper and steeper rate? That's the y-value, rocketing toward negative infinity. Don't follow it down there.

Of course, in the context of our problem, the sandwich stops when it hits the ground. But if there was no ground, it would keep falling and falling, down to negative infinity, like Gandalf. That's why this type of function is popularly known as a "McKellen function."*

So that is how you find a limit in Calculus. But will we ever discover a way to find our own limits?

3. Derivative: The rate of change at a specific point in a function.
And here we are, calculating rates of change, hoping to gain some control and lock things down by proving once and for all that things are changing and not locked down.

To keep our sanity, let's just apply this existential conundrum to our turkey sandwich "McKellen function."

$$f(x) = -4.9x^2 + 10x + 20$$

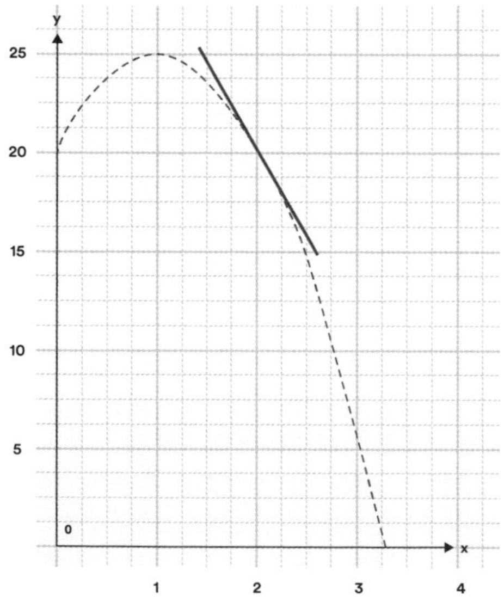

* NOT its real name

Let's take the derivative of the function at $x = 2$.

So right now, we are just looking for the rate of change at the point where $x = 2$, which can be measured as the slope of the blue line, which is *tangent* to the graph at $x = 2$. (In this context, *tangent* means the line touches the curve at only one place, without intersecting. It does not mean opposite/adjacent like it did in the Trigonometry chapter. Yes, the word "tangent" means two different things, like a parent saying to their kid, "You're brave to wear that."

The derivative of a function, usually written as $f'(x)$, is calculated with this relatable, easy-to-use formula:

$$f'(x) = \lim_{h \to 0} \frac{f(x+h) - f(x)}{h}$$

Let's try that out with our function, $f(x) = -4.9x^2 + 10x + 20$.

$$f'(x) = \lim_{h \to 0} \frac{-4.9(x+h)^2 + 10(x+h) + 20 - (-4.9x^2 + 10x + 20)}{h}$$

What?

$$f'(x) = \lim_{h \to 0} \frac{-4.9(x^2 + 2hx + h^2) + 10x + 10h + 20 + 4.9x^2 - 10x - 20}{h}$$

Um, if you say so.

$$f'(x) = \lim_{h \to 0} \frac{-4.9x^2 - 9.8hx - 4.9h^2 + 10x + 10h + 20 + 4.9x^2 - 10x - 20}{h}$$

Let's rearrange things:

$$f'(x) = \lim_{h \to 0} \frac{-9.8hx - 4.9h^2 + 10h - 4.9x^2 + 4.9x^2 + 10x - 10x + 20 - 20}{h}$$

Some parts cancel out, leaving us with:

$$f'(x) = \lim_{h \to 0} \frac{-9.8hx - 4.9h^2 + 10h}{h}$$

Which can simplify to:

$$f'(x) = \lim_{h \to 0} -9.8x - 4.9h + 10$$
$$f'(x) = -9.8x + 10$$

So the derivative of

$f(x) = -4.9x^2 + 10x + 20$ is:
$f'(x) = -9.8x + 10$

And the derivative specifically at $x = 2$ is:

$f'(2) = -9.8(2) + 10$
$f'(2) = -9.6$

The derivative, or rate of change, of this function when $x = 2$ is −9.6.

We see this on the blue line on the graph. See how the line travels horizontally from 1.5 to 2.5, a span of 1 second? During that time, the line travels down from (roughly) 25 to 15.4, or 9.6 meters in a negative direction. This is because the derivative at that point is −9.6.

This means that at exactly 2 seconds after you tossed your sandwich, it was traveling downward at 9.6 meters per second.

Whether you think Leibniz's work was derivative of Newton's, or Newton's work was derivative of Leibniz's, or neither, this is how it works.

> TL;DR: The main focus of Calculus is **to measure and predict change**. How does it work? By making sure you are so scared of how complex it is that you don't notice all the weapons it makes.

You must understand: The possibilities of Calculus, when taken to a high level, exceed humans' collective mental capacity, which may indicate that we shouldn't be messing with it too much.

What's in this chapter?

The problems in this chapter will ask you to do complicated, strenuous math, but your job is to see past that. Every problem has a straightforward, nonmathematical solution. You are smart enough and intuitive enough to figure out every single one of these problems. No Math needed. If you can't figure them out, who cares, this book is not a test.

Now, if you *are* some kind of rocket scientist, enjoy doing calculus, need to pass a class for your major, or have something to prove, feel free to take on the Math—it all adds up. We understand that it is often necessary to take lots of complicated steps in order to arrive at a simple solution.

#1. Ups and Downs

The last two months of Natalya's life are represented by this graph:

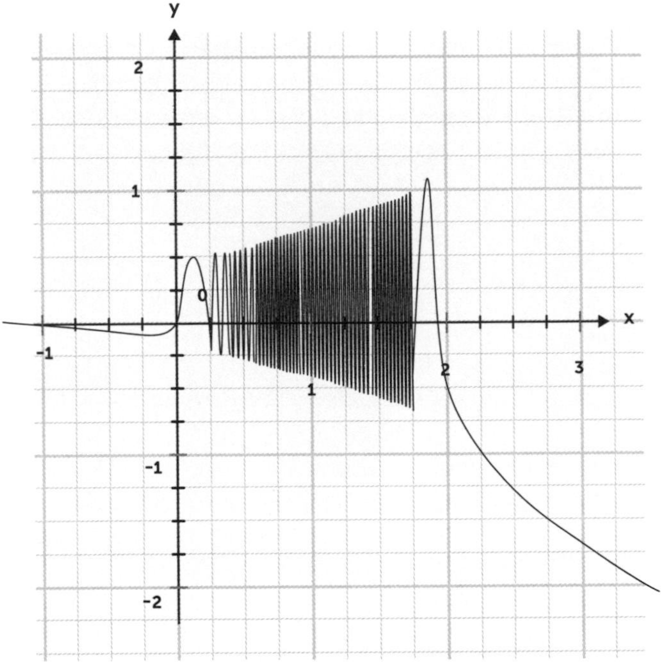

$$f(x) = .1351 - 2^{x-2} \cos\left(\frac{1}{(x-1)^{10}}\right)$$

Point (0,0) marks Natalya standing with Andy at the Gare du Nord train station, bag in her hand, beret on her head. "I'm sorry, Andy," she said. "I love you, but I have to make some changes if I'm going to reach my potential."

Look at the point on the graph that represents today ($x = 2$). Is the derivative positive or negative at that point? In other words, is Natalya on the mend today? Or on the decline?

TO SOLVE — Ups and Downs

There is no real problem here.

Emotional ups and downs after a major life change are normal. Natalya's mood is not something Math can measure. Let's move on.

Oh, you want to do the math anyway?

Um, alright. Then look at the graph and find the point where $x = 2$.

Once you've found it, determine if it is going up or down.

Pretty clearly going down.
There, the derivative is negative.
She's on a downward slope.
You proved it.
Happy now?

Are you the kind of person who says stuff like, "Hey, you look tired" to people you don't even know that well?

Bonus computations for anyone interested:

$$f(x) = .1351 - 2^{x-2} \cos\left(\frac{1}{(x-1)^{10}}\right)$$

$$f'(x) = -2^{x-2} \cos\left(\frac{1}{(x-1)^{10}}\right) \ln 2 - \frac{10 \sin\left(\frac{1}{(x-1)^{10}}\right) 2^{x-2}}{(x-1)^{11}}$$

$$f'(2) -8.789$$

#2. Know Your Limits

Your math teacher, Mr. Newman, is out today, and the substitute has drawn the graph of $y = \frac{1}{x}$ on the board:

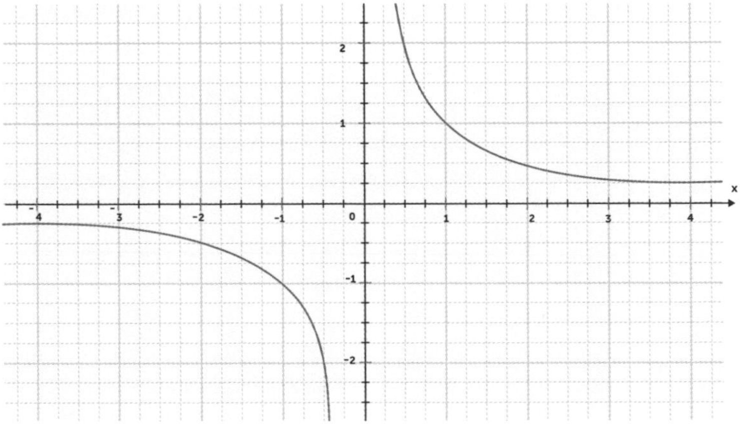

The sub explains that a “limit” is a certain y-value that a function approaches as it approaches a certain x-value, which is your cue to zone out and watch **McNiel** be a big shot.

But your rival’s annoying voice doesn’t pierce your eardrums like it usually does.

McNiel is just *sitting* there, shoulders slumped, gazing out the window.

Your rival’s book isn’t even open. They couldn’t know it was you who threw the rock, could they?

You take advantage of this moment to be the first to solve the problem.

A) What is $\lim_{x \to \infty} \frac{1}{x}$?

B) What is $\lim_{x \to -\infty} \frac{1}{x}$?

C) What is the limit of your desire to defeat **McNiel**?

TO SOLVE — Know Your Limits

C) This is the most important question. Where does this feud end?
Is **McNiel** a jerk? Sure, but you cast the first stone. Literally.
Two negatives
(Do not always)
Make a positive.

A) What is $\lim_{x \to \infty} \frac{1}{x}$?

Or: When the line along the *x*-axis gets to infinity, what is *y* going to be?

Let's look at the graph to understand it visually.

See how it's curving toward 0 on the *y*-axis? That tells us the answer:

$\lim_{x \to \infty} \frac{1}{x} = 0$, the number of feelings you thought **McNiel** had.

B) What is $\lim_{x \to -\infty} \frac{1}{x}$?

Let's look at a table to understand:

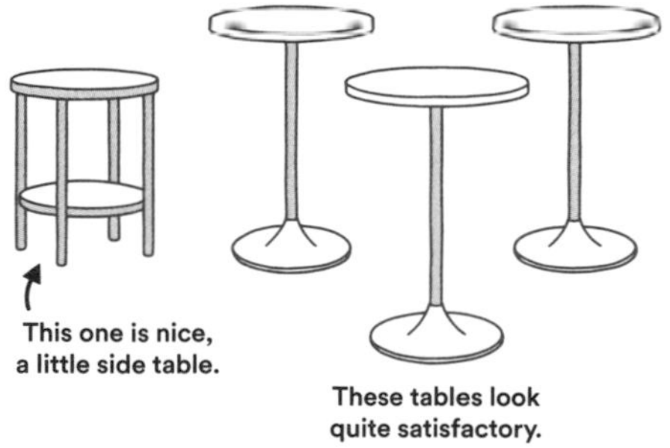

x	$y = \frac{1}{x}$
−1	−1
−10	−.1
−100	−.01
−1,000	−.001
−1,000,000	−.000001

So the $\lim_{x \to -\infty} \frac{1}{x} = 0$, what you feel like now, knowing you may have hurt **McNiel**'s feelings.

#3. Find the Subliminal Messages

$$\lim_{n\to\infty} \sum_{k=1}^{n} \frac{2}{n}[(\frac{2k}{n})^3 + 5(\frac{2k}{n})]$$

$$\lim_{n\to\infty} \sum_{x=1}^{s} \frac{2}{n}[\frac{h3y + y0u^3}{n^3} + \frac{10k}{n}]$$

$$\lim_{n\to\infty} \sum_{x=1}^{s} [\frac{Dr1nk^3}{n^4} + \frac{c0ke}{n^2}]$$

$$\lim_{n\to\infty} \sum_{x=1}^{s} [\frac{16}{n^4} \cdot k^3 + \frac{20}{n^2} \cdot \frac{n(n+1)}{2}]$$

$$\lim_{b\to\infty bs} [\frac{16}{n^4} \cdot (your\, fr13nds\, only \frac{hang\ 0ut}{with} you)^2 + \frac{20}{n^2} \cdot \frac{n(n+1)}{2}]$$

$$\lim_{n\to\infty} [\frac{16}{n^4} \cdot [\frac{n(n+j)]3sus^2}{saV3s} + \frac{20}{n^2} \cdot because\ they \frac{f3el}{s0rry} 4\, you \frac{n(n+1)}{2}]$$

$$\lim_{n\to\infty} [\frac{16}{4} \cdot \frac{[n(n+1)]^2}{n^4} + \frac{20}{2} \cdot \frac{(n+1)}{n^2}]$$

$$\lim_{n\to\infty} [\frac{aliens}{4} \cdot \frac{are}{n} \cdot \frac{real}{and} \cdot \frac{n+1}{they} \cdot \frac{n+1}{are} + \frac{20}{coming} \cdot \frac{no}{no} \cdot \frac{n+1}{no}]$$

$$\lim_{n\to\infty} [4\,(1 + \frac{1}{n})(m0m) - is - \frac{D1sappointed}{in^4 you} + 1(100k + th\frac{1}{n})]$$

$$4 + \frac{the\ government}{4} \cdot \frac{kn0ws}{n} \cdot \frac{about}{the} \cdot \frac{n+1}{porn} + 10$$

aliens

#4. Moonstruck

It's the last day of Virtual Reality Camp. Cody's got the second-highest score of the whole summer and he didn't even have to make any friends. If he aces this next task, he'll be crowned King of Camp.

Cody walks through the activity room, a dimly-lit warehouse full of pubescent boys and girls, all standing motionless. He puts on his sensory deprivation head mask and is instantly transported to a lunar landscape.

As he stands on the squishy surface of the moon, bow and arrow in hand, a space monster bears down on him from directly overhead. Teeth gnashing. Wings flapping.

Cody's got one shot to hit the target. He strings the arrow, pulls back with all his virtual strength, and lets it fly. The arrow takes the trajectory defined by $f(t) = 68t - .83t^2$. The arrow flies upward for 5 seconds before it hits the space monster in the center of its largest eye. The beast drops dead, falling slowly because they're on the moon.

CONGRATULATIONS, KING OF CAMP!

Cody is ecstatic: "Wait until Dad hears I'm a champion!" He jumps for joy on the squishy moon, as well as in real life. What Cody cannot see is the rock from page 61. It rolls aimlessly into the room. It knows nothing of kings or champions.

Landing on the rock in reality, Cody hears a loud snap in his ankle and, after the shock, feels the most intense pain he's ever known. He cries for help, but no one can hear him through their sensory deprivation head masks.

How fast was the arrow rising when it struck the space monster?

#5. Please, Have a Seat

Natalya's going to the movies—she doesn't want to miss the latest summer blockbuster, *Space Heroes vs. The Government*.

She wants to maximize her viewing angle.

The screen is 20 feet high, and the bottom of the screen is 10 feet from the floor. The first row is 8 feet from the wall, and each row is 2 feet behind the row in front of it.

While looking for the perfect seat, Natalya hears an all-too-familiar laugh.

Andy,
In the 4th row,
Eating popcorn,
With his blue-haired date.

Natalya's 1 task immediately becomes 4 tasks:

1. Remember how to breathe.

2. Pick up the Raisinets she's spilled on the floor.

3. Find a seat with a good viewing angle.

4. Remain out of the field of vision of Andy, who once held her so close.

Where should Natalya sit?

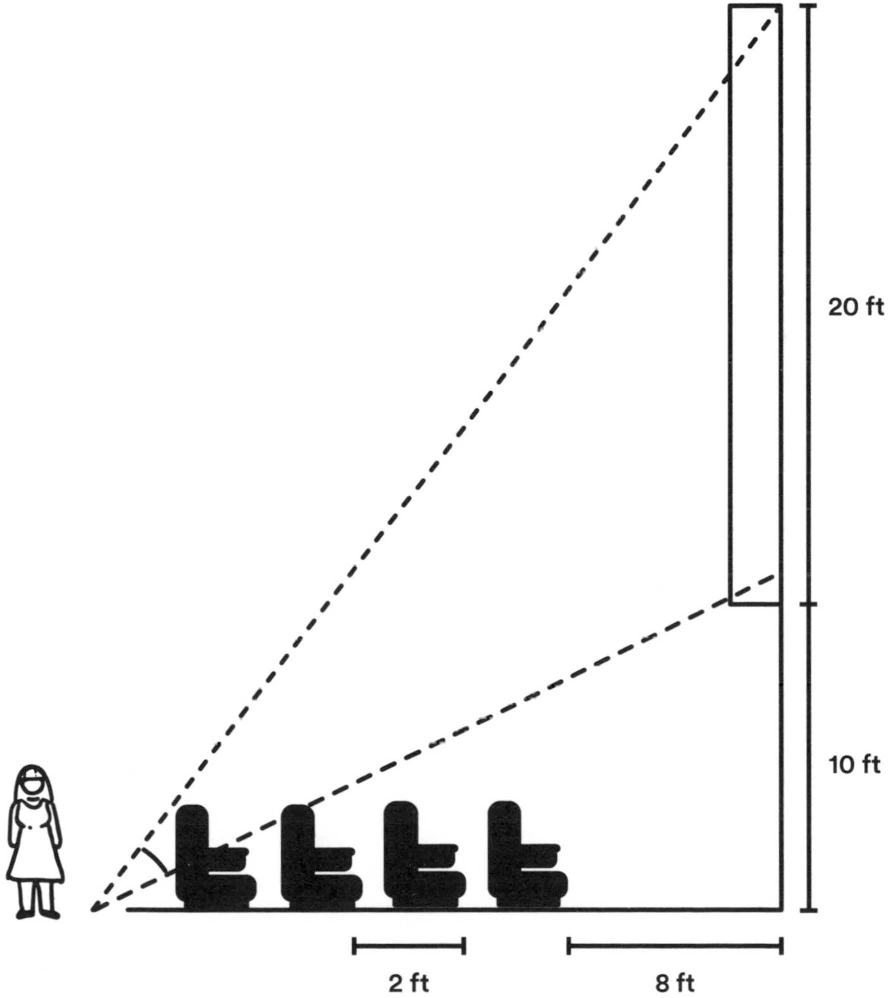
20 ft
10 ft
2 ft
8 ft

FACTS & OPINIONS (FROM MIKE)
SPACE TRAVEL

Space. The final frontier.

Because Calculus can be used to link an object's position, velocity, and acceleration at any given moment, it is useful in almost all aspects of space travel, including determining how and when exactly to launch a rocket in order to achieve a desired trajectory.

In 1957, the first dog orbited Earth. Her name was Laika, and her journey was a major stepping stone in space travel, as she proved that a mammal could survive outside of the atmosphere.

Ever since, we have only moved forward. In 1969, the first humans walked on the moon. "One small step for man," said Neil Armstrong, "one giant leap for mankind." The two other guys apparently said nothing, and just like that, another milestone was reached.

The United States spends about $19 billion a year on space travel and exploration, devoting funds to:

- Space Launch System (heavy-lift launch vehicle, basically an updated version of a space shuttle)
- Orion Spacecraft (multipurpose vehicle designed to facilitate human exploration of asteroids or Mars)
- Training and education for astronauts

These days, space travel prospects look better than ever. In addition to NASA's ongoing pursuits, space travel has expanded into the private sector. SpaceX's Elon Musk is now a household name, and his company's work on spacecrafts that can travel to Mars gives hope to millions.

For millennia, outer space has seemed to be beyond us. But with Calculus at our disposal, even the distant expanses of the galaxy may now be within our reach.

OPINIONS & FACTS (FROM TAYLOR MARIE)
SPACE TRAVEL

Space. The final frontier.

FINAL, meaning it's the *last* thing we should address.

Calculus has made us too smart for our own good, giving us the tools to explore space. We can now get in a big metal tube and shoot ourselves up into the sky, out past the ozone layer and as far away from our problems as possible.

In 1957, Laika, a stray dog from Russia, was launched into space. And she died up there. Have you ever seen a dog locked in a car with the windows rolled up? This was that, times a million.

Ever since, we have been sending people and robots and music and fancy cars into space, just to float about and see what's going on up there. Meanwhile, the ocean takes up seventy-one percent of Earth's surface and we've only explored five percent.

The United States spends about $19 billion a year on space travel and exploration, money that could be spent on any of the following:

- Education ($5,000 raise for every teacher)
- Mental health research
- Sustainable farming
- A nice electric toothbrush for every American, or one printer ink cartridge. Your choice!

These days, space travel prospects rest in the hands of billionaires. Elon Musk works tirelessly to send humans to Mars instead of focusing his efforts on sustainable energy, something everyone would benefit from.

For millennia, outer space has seemed beyond us. But with Calculus at our disposal, we can reach farther than ever before, hoping that whatever we find will somehow make us feel whole.

#6. Excuse Me, I Need to Take This Call

The sun is shining on Kurt Midas, president and CEO of the Kurt Midas Oil Corporation. He's opening a new oil well, feeling comfortable surrounded by a group of men just like him. He'd feel more comfortable if he knew he had more money than them, but there's still time for that.

He excuses himself to take a call.

It's not great.

> "Sir! One of your wells just sprang a leak and now thousands of gallons of oil are spreading out in a circular shape in the middle of the Gulf of Mexico! The area of this circle is increasing at a constant rate of 10 square feet per second! What should we do?"
>
> He hangs up the call,
> ignores another call,
> picks up the big scissors,
> and cuts the red ribbon.

The men applaud, then laugh, then compliment each other in both backhanded and fronthanded ways.

How fast (ft/sec) is the radius of the circle increasing when the oil spill measures 200 miles across?

#7. An Apple a Day is Not Sustainable in This Economy

Anita wanted to keep her farm in the family, but when she settled in to do the math, it was clear from her dwindling profits that time was running out. She has children to think of. Anita figures she can use the money from the oil corporation to buy some low-risk stock, or that bitcoin stuff.

Packing up her possessions, she stares out the window at her apple trees. *Trees are timeless, mighty, and tragic figures,* she thinks, folding her work pants. *You never know exactly how much wisdom a tree has until you chop it down and count its rings. Perhaps there is a metaphor in there. But for what, I do not know.*

Anita decides to put on her work pants and plant some new trees, one last time. This year, Anita's 100 trees each produced 60 apples. She knows that for each tree added to the orchard, the apple production per tree decreases by 2 apples.

If Anita wants to grow as many apples as possible, how many trees should she plant?

#8. Edible Arrangements

Match these images to the food they remind you of:

A) 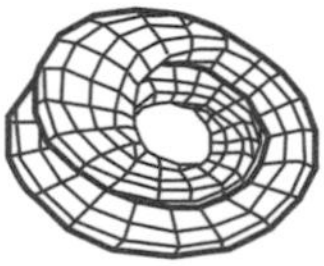

ICE CREAM CONE

B) 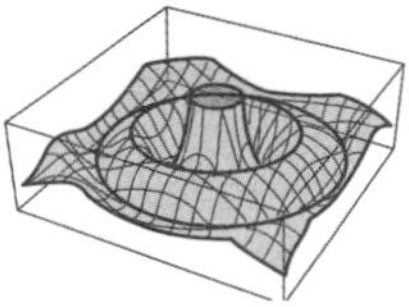

CARNITAS TACO

C) 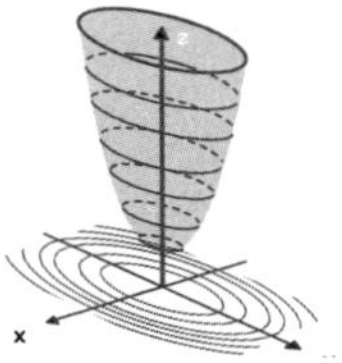

BOWL FOR CHIPS & DIP

D) 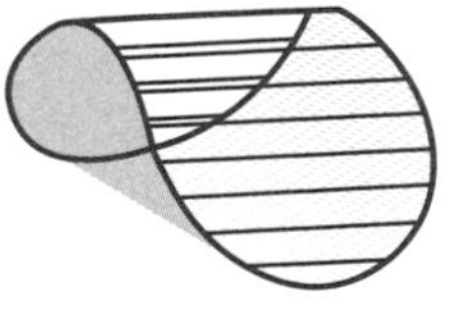

CINNAMON ROLL

Pick one to have as a treat—you've been doing a lot of hard work!

#9. 'Til Death Do We Drop

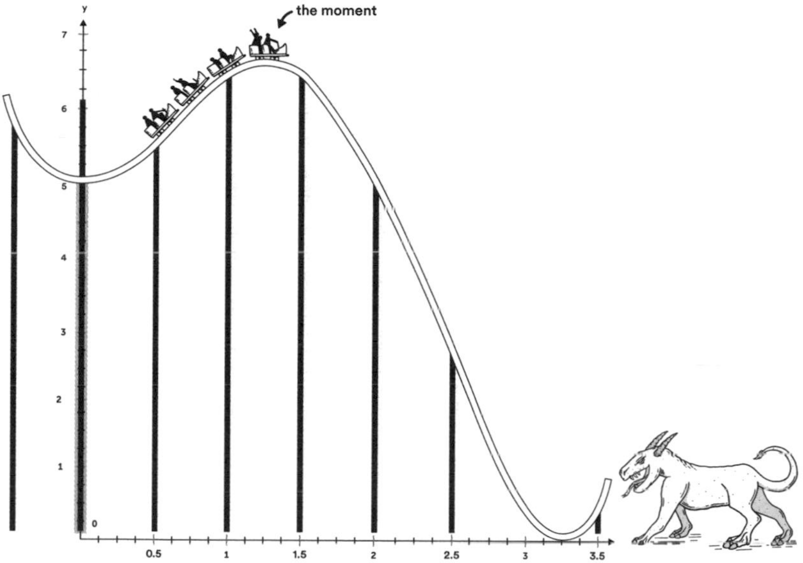

The above graph shows the function of $f(x) = \frac{1}{2}x^4 - 3x^3 + 4x^2 + 5$.

It also shows Hellbeast's Surprise, the biggest roller coaster in the state.

At the coaster's peak, Bart takes out an engagement ring, a hollow circle that cost him three months' salary. He wants Sonya to know how much she means to him, but has forgotten about her debilitating fear of heights. As she screams, "PLEASE, GOD, DON'T LET THIS BE THE END," Sonya accidently knocks the ring out of Bart's hands. They watch the hollow circle fall down, down, down into the Hellbeast's mouth, never to be seen again.

A) If an engagement ring is the truest measure of love, then exactly how much does Bart love Sonya? (Hint: We know Bart's salary from a previous problem.)

B) At the moment of the proposal, what is the derivative of the roller coaster?

CHAPTER SUMMARY

You have slain the beast! The Calculus demon slinks back down into the dark fire from whence it came, the hellmouth snaps shut, and you stand victorious, tasty treat in hand.

1. A limit is the specific y-value that a function approaches as it approaches a specific x-value.

2. Calculus was established by two separate people at the same time, but white men, notoriously bad at sharing, refused to accept this.

3. Emotional ups and downs are normal after a breakup.

4. In Math, two negatives make a positive, but in life, they do not.

5. Knowing Calculus doesn't necessarily mean you are smart, and not knowing Calculus doesn't necessarily mean you are not smart, but it is smart to know that Calculus isn't necessary.

6. Trees are timeless, mighty, and tragic figures.

REVIEW QUESTIONS

1. Would you go see the third movie in the trilogy, *Space Heroes: Back to Before*?

2. What about the fourth, *Space Heroes: Intergenerational Drama*?

3. How interested are you in the animated TV show prequel, *Space Heroes Fight Evil: Robot Altercations*?

4. Would you play the video game, *Space Heroes: Melee Smackdown*?

5. What about the other video game, *Space Heroes vs. NBA Legends*?

6. How likely are you to see the spin-off film, *Space Heroes Help the Water Aliens, Remember Them from Before*?

7. What about the sequel to that one, *Space Heroes vs. Water Aliens: The Battle for Forever*?

Thank you for participating in this market survey.

Chapter Five

PROBABILITY & STATISTICS

CALCULATED RISK TAKING, AND HOW TO COMPARE YOURSELF TO OTHERS

Hi, and welcome to Probability and Statistics, the math class with the most practical applications and the fewest students in weekly psychotherapy. This is the kind of Math that is most useful. If you skipped Calculus and went straight here, good job knowing yourself.

Where did it come from?

The concept of **Probability** has been known to humankind for as long as games of chance have been played, making it the oldest type of Math covered in this book. As long as there have been playing cards (more than 1,000 years), six-sided dice (more than 5,000 years), and insurance (almost 4,000 years), probability has been here, too, helping people predict the future.

Statistics is a fancy word for the collection and manipulation of information. For most of history, statistics mostly focused on material resources, helping answer questions like "How much can I tax this stuff?" and "How many people in this village are able to fight in my big scary war against those *different* guys from the other village?"

Nowadays, Probability and Statistics are used for lots of other fun things, like predicting your behavior and tracking your interests so companies can sell you clothes, prescription medications, and ideologies.

How does it work?

Probability is a measure of likelihood, rated on a scale from 0 (impossible) to 1 (definite). If you flip a coin, the probability of it landing heads up is $\frac{1}{2}$. If you wake up in a Dumpster with no idea how you got there, the probability of it being Tuesday is $\frac{1}{7}$. The probability that it isn't your girlfriend's birthday today is $\frac{364}{365}$ (unless it's a leap year, then it's $\frac{365}{366}$). See how it works?

Probability inherently denies the concept of "fate," assuming that our universe is not governed by any forces we don't yet understand. It puts its faith in the ability of past information to predict future information: calculated risk taking. But does calculating a risk teach you as much as actually going out and taking that risk? According to the popular film *Along Came Polly*, no.

Statistics is the mathematics of presenting and understanding data. Whenever you see a graph or chart, you are looking at statistics. Stats people like to throw around a lot of fancy terms like "standard deviation," "cost-benefit analysis," and "quarterback rating." In this chapter, we're just going to focus on three: **average**, **median**, and **mode**.

AVERAGE

Calculated simply by $\frac{\text{sum of the parts}}{\text{number of parts}}$

An average is also called a "mean" because condensing a lot of complex information from different people living in different circumstances into one number is exactly that: mean. In fact, averages are not only mean, but potentially misleading. Let's say you respond to an ad that offers you a place to live. "The average age of the 5 people who live in this house is 30," claims the ad. Great, you're about 30, you'll fit right in. But when you show up, you are greeted by 4 18-year-olds and an 80-year-old. The average age is about 30, yes, because:

$$\frac{18 + 18 + 18 + 18 + 80}{5} = 30.4$$

However, by no means does this piece of data accurately reflect whatever the hell is going on in this house.

MEDIAN

Calculated by selecting the middle number.
For this house, the median age is 18, because:

18 18 18 18 80
~~18~~ ~~18~~ 18 ~~18~~ ~~80~~

Probably more accurate to the situation, but makes no mention of the outlier.

MODE

Calculated by the number that appears most frequently.

In this situation, the mode is the same as the median: 18.

Still . . . what's the octogenarian doing there? Are they someone's relative? Condensing information using Probability and Statistics can help us see things with more accuracy, but it doesn't always help us see the truth.

Here's a tip to understand the difference between the two concepts: Probability is seeing a student and predicting their test scores; Statistics is seeing the test scores and guessing which student they belong to.

What's in this chapter?

The problems in this section see our characters in situations in which they must engage in calculated risk taking, try to understand graphs and charts, and/or apply statistical data to their own lives. Where do they fit in? Will their behavior change once they know more information? What does the data say about who they are and what they do? Are they average? Are they special? Is it possible to be both? Probably, and statistically speaking, no.

#1. In the Golden Age of Television

In the United States, the top 11 most viewed TV shows are as follows:

PROGRAM	AVG WEEKLY VIEWERSHIP
White Men Decide What's News	18,200,000
Particularly Nasty Crimes, and the Response to Those Crimes	16,800,000
Counting Rocks with Professor Don Obvious	16,100,000
Cops with Personal Problems	15,650,000
Too Many Women in One House	14,900,000
Doctors Cheat on Their Spouses	13,500,000
Sing for Your Supper!	13,500,000
That Show from Years Ago, but It's Now	12,400,000
Former Athletes Discuss Current Athletes	12,300,000
Celebrity Lip-Sync Pie-Eating Contest	12,100,000
Bearded Fantasy Sex Fiesta	11,800,000

A) What is the average weekly viewership of the top 3 programs?
B) What is the median weekly viewership of the top 5 programs?
C) What is the mode weekly viewership of the whole chart?
D) If this is the Golden Age of Television, why are most actors unemployed?

TO SOLVE — In the Golden Age of Television

A) The average of those 3 programs will be:

$$\frac{(18{,}200{,}000 + 16{,}800{,}000 + 16{,}100{,}000)}{3} = \frac{51{,}100{,}000}{3} \approx 17{,}033{,}333 \text{ viewers}$$

But if you *really* want to be average: Develop a predictable weekly schedule, don't wear any flashy statement pieces, don't have any loud opinions, and watch all these TV shows with a group of nonintimidating, unremarkable friends.

B) The median of those 5 programs will be:

14,900,000	15,650,000	16,100,000	16,800,000	18,200,000
~~14,900,000~~	~~15,650,000~~	16,100,000	~~16,800,000~~	~~18,200,000~~

16,100,000 viewers.

C) The mode is the most common number, which on this table is 13,500,000.

D) The market is oversaturated—there are more applicants than jobs available. And most aren't truly unemployed, they're waiting your tables and walking your dogs. Please tip them 20%.

#2. Mind the Gap

The average American man makes $1.26 to every dollar that an American woman makes for the same amount of work.

Do you think this has something to do with why some women have a negative experience with Math?

#3. Dress for Success

Andy's got a date tonight with Lea, the blue-haired girl from the movies. They're going to Smarter Child, that new bar downtown with the arcade games. It's a popular hangout for millennials who feel nostalgic about decades they didn't experience.

Andy feels like he's close to that healthy relationship he's been looking for. He just needs to get dressed. He has a limited supply of clean clothes.

SHOES	SHIRTS	PANTS
Tennis shoes	Blue collared shirt	Black pants
Brown shoes	Gray T-shirt	Blue jeans
Dress shoes	Hawaiian print shirt	Cargo shorts
Scuba diving flippers	That sweater Natalya bought him in Paris	

How many outfit options does Andy have?

TO SOLVE — Dress for Success

To calculate this, we need a spot for every choice he has to make:

____ × ____ × ____

In the blanks, fill in the number of options he has for each choice:

$\underline{\ 4\ } \times \underline{\ 4\ } \times \underline{\ 3\ }$

He has 48 options.

HOWEVER...

If he wears the flippers, he'll project a major lack of social awareness, and won't be able to play *Dance Dance Revolution*.

If he wears the blue collared shirt, he shouldn't wear the dress shoes. She might think he's taking this too seriously.

If he wears the Hawaiian shirt, he shouldn't wear the cargo shorts. She might think he's not taking this seriously enough.

If he wears the sweater Natalya bought him in Paris, Lea might be like, "Hey, that's a nice sweater, where'd you get it?"

If he really feels comfortable around her,
he should be able to wear whatever he wants.

#4. Butts

Even though Natalya didn't get funding for her company, a nonprofit that turns cigarette butts into school supplies for children, she walked around all day collecting cigarette butts from the ground, because she believes in her dreams.

Natalya picked up 300 cigarette butts from 4 different brands, some with lipstick (represented in blue), some without (represented in gray):

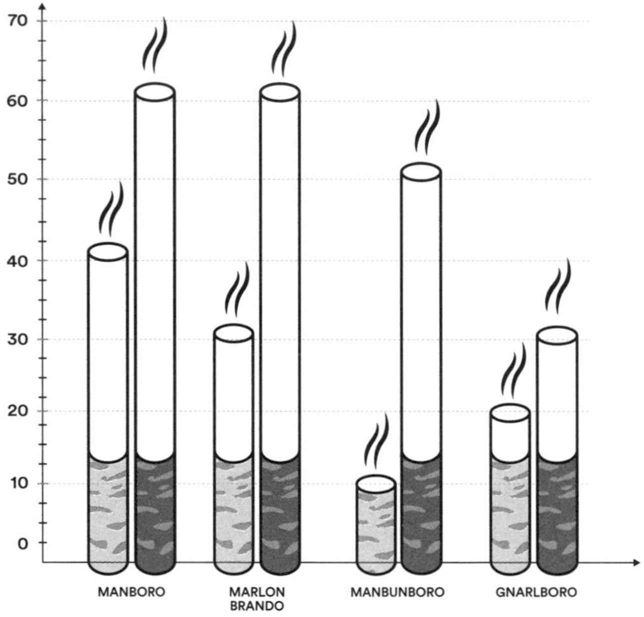

Natalya sets down her bag in the workshop between her refrigerator and the toilet. She picks up one cigarette at random. "Here we go," she says. "Time to make the world a better place for the children."

A) What is the probability that this cigarette is a Manbunboro?
B) What is the probability that this cigarette has lipstick on it?
C) What is the probability that this cigarette is a Marlonbrando with lipstick?
D) What is the probability that this company will be successful?

#5. Shoot

Mr. Newman is still out, and for some reason the substitute thinks it's OK to give you a pop quiz. It reads:

YEAR	LOCATION	FATALITIES	GENDER
2017	Las Vegas, NV, music festival	59	Male
2016	Orlando, FL, nightclub	50	Male
2007	Blacksburg, VA, college	33	Male
2012	Newtown, CT, elementary school	28	Male
2017	Sutherland Springs, TX, church	27	Male
1991	Killeen, TX, restaurant	24	Male
1984	San Diego, CA, restaurant	22	Male
1966	Austin, TX, college	18	Male
2017	Parkland, FL, high school	17	Male

These are the deadliest mass shootings in the United States since 1950.

A) What is the common factor?
B) How many of these happened between 1950 and 1969?
C) Between 1970 and 1989?
D) Between 1990 and 1999?
E) Between 2000 and 2009?
F) 2010 and after?
G) What is the probability that if all these mass shootings were perpetrated by women, lawmakers would handle the situation differently?

#6. Château Hellaux

Bart is having dinner at Château Hellaux with his bosses, Mr. Ball and Mr. Shaft, who are looking to promote someone to partner.

Dinner is great. Bart has a winning attitude, he has all the right qualifications, and he had the porterhouse steak. It was . . . disappointing.

The bill comes to $150.67. Mr. Ball hands the check to Bart to calculate how much they should tip. (Of course, he wouldn't do the math himself. Now that he's an executive, he lets others handle his busy work.)

The waitress seemed annoyed. Mr. Ball was talking on his cell phone when she came to take their order, and Mr. Shaft took 10 minutes deciding between the 1974 Petrus and the 1975 Petrus before ordering a martini. He also kept calling her "sweetheart" and asking questions like "Do you have any more tattoos we could see?"

How much should Bart tip the waitress? Whose name, by the way, is Michelle.

TO SOLVE — Château Hellaux

To find a percent, you must multiply the total amount by the *decimal version* of the percent you want.

OK—that means:

If you want to find 50%, multiply by .50

If you want to find 25%, multiply by .25

You want to find 1%? Multiply by .01

Or if you want to find *the* 1%, go to Wall Street. (Hint: They're the ones giving themselves million dollar bonuses.)

A 20% tip on $150.67 is: .2 × $150.67 = $30.13

But Michelle had to deal with quite a lot of Ball and Shaft tonight—more than she expected, or wanted. Let's tip her 25% at least. She has a big audition coming up.

.25 × $150.67 = $37.67

Now round up to an even $40. That'll make the addition really easy when you calculate the total for the bill. Don't round down; it's rude.

Here's another good tip: If you see your buddy treating a woman with disrespect, you can say something to him, instead of just letting it happen and then paying her after. Or you can do both. You know what? Do both.

FACTS & OPINIONS (FROM MIKE)
PERCENTAGES

Percentages are one of the most useful pieces of Math. They are used to describe specific amounts, or parts of a whole.

Percentages are commonplace in our everyday lives:

- 50% chance of rain
- 25% off these jeans
- 5% APR financing

Percentages are useful for calculating interest and balancing a diet, and are the cornerstone of an effective government: taxes.

Let's go back thousands of years: In the Bible, Joseph explains to the people of Egypt that they must give twenty percent of their crops to Pharaoh. In ancient Rome, taxes were collected by a publican—a private citizen who had successfully bid for the right to collect taxes on a certain area.

Percentages are here to help us compare things with one another. Whether you are an investment banker, a sports fan, a salesperson, a gambler, or you just like to eat at restaurants, percentages have always, and will always, play a role in your life.

OPINIONS & FACTS (FROM TAYLOR MARIE)
PERCENTAGES

Percentages are one of the most overvalued pieces of Math. They are used to describe specific amounts, or parts of a whole.

Percentages shout in our faces on a daily basis:

- 11% battery remaining!
- 99% accurate pregnancy test!
- 2% milk! (what the other 98% is, we may never know)

Percentages are useful for fun things like seeing how much lead is in your water, calculating body fat, and every child's favorite way to measure themselves against their peers: grades.

Let's go back to the fourth grade for a moment: "Hey, Mom, let me get this straight: I'm not supposed to compare myself to the other kids at school, but I *am* supposed to do my math homework so I don't fall behind the other kids at school?" What the hell is going on here?

Percentages help us compare things with one another: rich and poor, men and women, black and white, fat and skinny, young and old, you name it! So thanks a lot, percentages, for helping us focus on parts of the whole, instead of the whole picture.

#7. Numerology

Take the numerical value of the month that you were born and add that to the day that you were born.

Add that sum to the year that you were born.

Divide that number by your current age.

Take that number and multiply it by the number of letters in your first pet's name.

Now take the square root of that number, add your favorite number, then subtract your least favorite number.

Write that sum on a piece of paper, put it in your mouth, chew it and swallow it.

There, did you learn anything about yourself?

#8. Aliens Exist

A tour guide hovers past a glass case and welcomes a tour group to the Hall of Machines, which features the Minivan Habitat, the ThighMaster Exhibit, and now the Human Display.

Your math teacher, Mr. Newman, is the newest exhibit at the Trappings of Earth Museum. Time seems to run differently on the planet DeLonge, and he has no idea how long he's been away from home. He wanted to seem distant, but not this distant.

He has given up on trying to explain to his captors the difference between human and machine: how the ThighMaster can never know love, and how a minivan driving down a highway can never undersand just how free that can make someone feel.

He misses his home, his car, his students, a cup of black coffee in his favorite mug. The alien coffee is blue and salty. He never understood until now how lucky he is to be a human—to understand both numbers and emotions.

As the tour guide beeps and gurgles, Mr. Newman silently flips a penny. It lands on heads 6 times in a row.

The alien lets out a high-pitched scream, and its many tentacles start waving wildly, tying themselves into knots. Apparently, the number 6 is extremely bad luck on the planet DeLonge. Before he knows it, Mr. Newman is forcibly removed from his glass case, pushed into a small pod, and launched through space toward Earth.

What were the chances of Mr. Newman flipping 6 heads in a row?

TO SOLVE — Aliens Exist

To calculate the probability of a **sequence of events,** first find the probability of each individual event occurring. Then multiply them (you can use fractions or decimals).

Each time, the probability of the coin landing heads is 50%, or $\frac{1}{2}$.

Only two options here, assuming the coin doesn't land on its side, or disappear into thin air, or turn into an eye and give a flirty wink.

The probability of a coin landing on heads 6 times in a row is:

$$\frac{1}{2} \times \frac{1}{2} \times \frac{1}{2} \times \frac{1}{2} \times \frac{1}{2} \times \frac{1}{2} = \frac{1}{64} = .015625 = 1.5625\%$$

or

$$.5 \times .5 \times .5 \times .5 \times .5 \times .5 = .015625 = 1.5625\%$$

#9. Seating Chart

Bart and Sonya are making the seating chart for their wedding. Bart got that promotion at Ball, Shaft, & Associates, so they have some extra money to spend. They'll probably get a chocolate fountain, and rent some sort of amusement for the children.

A) They have 42 seats. How many ways can they arrange their 42 guests?

B) If it costs approximately $80 per guest based on food and alcohol alone, how much money could they save for something practical if they just stopped all this stressing right now and canceled the wedding reception?

C) If a wedding is an agreement to spend the rest of eternity with just each other, then why do they need to pay for all these other people to be there?

#10. Piece of the Pie

Anita and Kurt Midas are on separate planes heading to the big city, both working on their budgets. The first pie chart below represents Anita's plan for how to spend the $1,200,000 she earned from selling her land. The second pie chart represents Kurt Midas's plan for how to spend the first $1,200,000 he earns off that land.

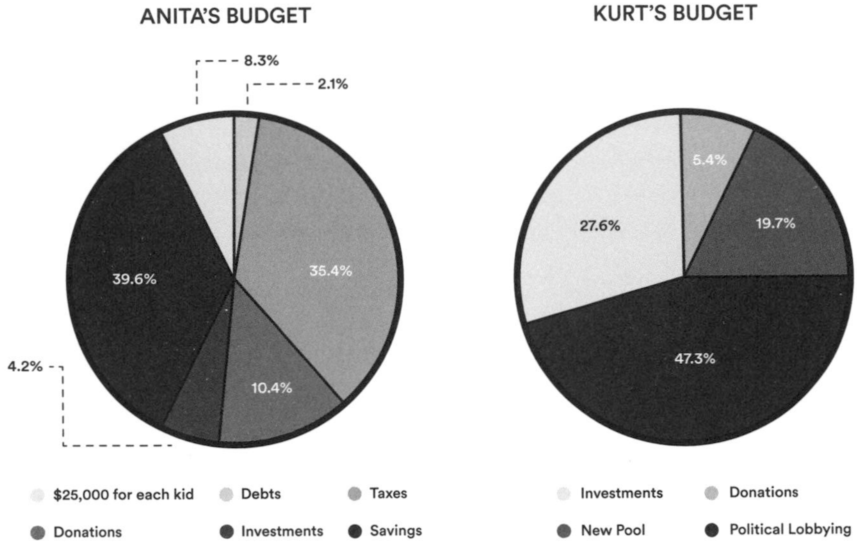

A) How much is she paying in taxes?
B) How much is she donating?
C) How much is she saving?
D) How much is he paying in taxes?
E) How much is he "donating"?
F) How much is he saving?
G) How is any of this fair?

#11. Deer Population

The rock from page 61 rolls through the forest,
caring for none,
wanting for naught.

It passes some trees, a pond, and a serene herd of deer before rolling over a crumpled piece of paper with the following chart on it:

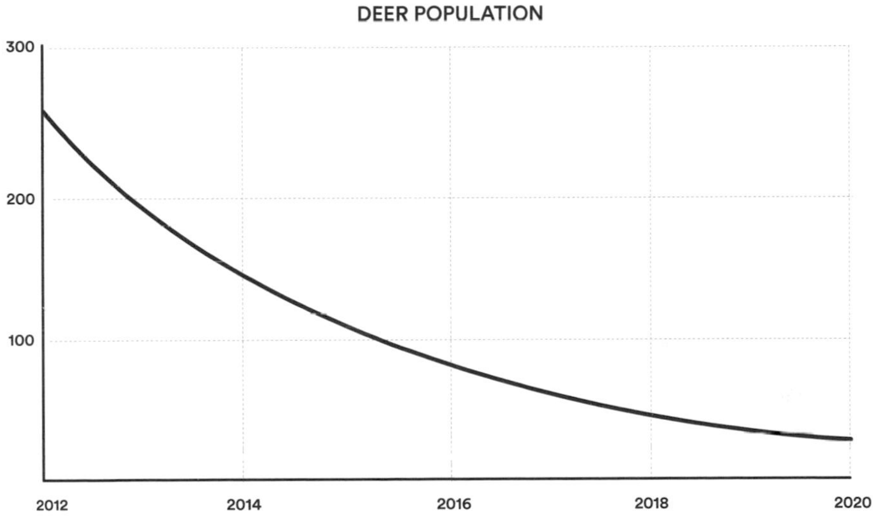

$$P = 256(.75)^{(x-2012)}$$

A) What will the deer population be in 2020?

B) What will the deer population be in 2031?

C) What will the deer population be today if a large stag, who embodies the ancient majesty of the forest, eats the paper?

#12. Referendum 18.C.2.D

The city is voting on a referendum that would establish a .3% income tax increase for this year in order to construct and maintain the Traffic Nightmare Coliseum, a new stadium for the tackle football team that represents the hopes and dreams of the community.

A) If the mean income in the city is $46,372, and the population is 3 million, what will be the total amount raised for the Traffic Nightmare Coliseum?

B) If Kurt Midas, the team owner, earns $125 million a year from his oil business, and has an estimated net worth of $2.3 billion, then why doesn't he just pay for it?

#13. This One Is About You and McNiel

You just got your most recent report card, and to your surprise you've done quite well. You peeked over at **McNiel**'s desk, just as you do every week, and caught a glimpse of your rival's grade.

The following chart shows your grades and **McNiel**'s grades over the last 4 weeks:

	YOU	McNIEL
WEEK 1	60	100
WEEK 2	68	98
WEEK 3	73	97
WEEK 4	88	88

You want to gloat, but after class **McNiel** is nowhere to be found, so you hop on your bike and head to your rival's house, trying to think of the perfect put-down.

"We're the same, you and I!" you'll say, finger pointed in **McNiel**'s face. "The math says so!"

You hop off your bike while it's still moving and strut up to the front door, but you stop short; there's shouting inside.

"B+ in math won't cut it in *this* house!"

"Dean Anantha P. Chandrakasan at MIT does not care that you can do a 10-second wheelie!"

You try to leave quietly but trip over your bike, knock over the garbage, and step on a cat. **McNiel** comes outside to investigate the commotion.

"Oh," says your enemy. "**McEvoy**, it's you."

"Yes, it's me," you respond, covered in trash. "**McEvoy**."

You, **McEvoy**, peel the old slice of pizza from your face and discover that you're blind. Wait, no, that's just two pepperoni slices stuck to your eyes. You take those off and you can see clearly: It's a familiar expression on your rival's face—the look of someone who has been bullied. You feel something inside. Like daytime cold medicine, but without the aftertaste. This feeling, dear friend, is compassion.

"Hey, I didn't know you could do a 10-second wheelie. Pretty cool. I can only do a 7-second wheelie."

McNiel smiles, extends a hand, and pulls you out of the garbage.

"I'm sorry I stepped on your cat."
"I'm sorry I stole your math book."
"I'm sorry I broke your window."
"You broke my window?"
"... No."

A) On average, how much did your grade increase per week?

B) Draw a line graph of your grades and **McNiel**'s grades.

C) Send a nice email to an old friend.

#14. I Crumble for Ya

Natalya needs a date for Sonya's wedding. If she shows up alone, Sonya's mom will start recommending cat adoption places again, and Natalya will have to explain that she has no time to take care of a cat—she's busy working on a nonprofit that turns cigarette butts into school supplies for children.

Here's her first Crumble match:

PASCAL, 29
ENGINEER

- Music
- Coffee
- Roof-top Dance Parties
- Cold Beer
- Travel
- Secret Goofball ¯_(ツ)_/¯
- Dog Person
- Proud Vegan

For any Crumble match . . .

- There is a 50% chance of the person responding
- From there, a 40% chance of setting up a date
- From there, a 70% chance of them showing up for said date
- From there, a 20% chance of a real connection
- From there, a 64% chance of them having HPV

What are the chances of Pascal responding, setting up a date, showing up, genuinely connecting with Natalya, and not having HPV?

#15. The Grand Hypatia

You and your best friend, **McNiel**, are riding your bikes downtown. You're headed to the library to loiter and pop sick wheelies in the parking lot.

Suddenly you slam on the brakes. A man you recognize from TV is yammering into his phone, walking across the street like nothing in the world could ever stop him. You find it odd he doesn't help two people fumbling behind him: a woman juggling presents, and a boy on crutches.

Now, more people are crossing the street. They're heading into the Grand Hypatia Hotel:

- A weathered woman and her three daughters
- A blue-haired girl and her date
- A couple of big shots
- A woman who smells like cigarettes and her date

And us.

We arrive with 30 seconds to spare—we've been very busy writing this book, you see. Struck by our effortless gravitas and approachability, you ask, "Hey, what's going on in there?"

We whip out forged invitations for you and **McNiel**, and you follow us inside.

What, did you think we'd miss the wedding?

CHAPTER SUMMARY

You've made it to the end of another chapter—what were the odds of that? Let's go over what we covered this time:

1. Probability is used to predict your behavior. Statistics are used to analyze your behavior.
2. Median and average can both represent sets of data, but neither one shows the whole truth.
3. You can calculate the probability of a sequence of events by multiplying the probabilities of each individual event.
4. Compassion is the only force in the universe that can turn an enemy into a friend.
5. Your name is **McEvoy**.

REVIEW QUESTIONS

1. What is the probability of rolling snake eyes (1 and 1) on a pair of dice?
2. Find the median, the average, and the mode of this series of numbers. Which gives the most accurate read of the information?

   3 4 5 5 5 7 8 10 10 11 11 13 17 20 25 27 27 40 59 69 89

3. When was the last time you got an STD test? Do you need to go in for one?
4. What percentage of all United States presidents are women?
5. What percentage of United States citizens were born unto a woman, who carried them inside her body for 9 months, feeding them from the inside and literally building them from scratch?

Chapter Six

THE WEDDING

OR, WHO CARES ABOUT BART & SONYA, ANYWAY?

Where did it come from?

In ancient days, weddings were unceremonious: Ugth the caveman, after attacking his neighboring village, grabbed woman, threw her screaming over his back, and went home. The spirit of these early weddings lives on, but like most things human beings touch, they have become far more complicated over time.

The decades-long follow-up to a wedding is known as "marriage," "matrimony," or the grim "wedlock." In medieval times, marriage was a way to strengthen alliances and family bonds, with a family marrying off their human daughter in order to secure money, land, and/or social status. She would then smile, cook, and bear children for her husband, who was often two or three times her age. Of course, she didn't know that, because no one taught her Math.

How does it work?

Weddings typically work like this: Two human beings invite their friends and family to take a day or two off of work, get dressed up, and travel to a scenic location or big room somewhere to watch the couple commit to each other forever. (In this context, the mathematical definition of "forever" is pretty flexible.)

In preparation for this major rite of passage into "forever," the couple goes through several minor rites of passage, which may include having dinner with all their parents, creating a wedding website, and/or dissecting a chicken and inspecting its liver to make sure it's healthy.

Sometimes, after the wedding, the couple goes on a special vacation called a "honeymoon." They often miss several days of work for this, because as we know, once you bring the moon into it, you don't have to use logic anymore.

How does the old nursery rhyme go?

First comes love,
Then comes marriage,
Then comes the baby in the baby carriage,
Then comes a divorce
for forty to fifty percent of couples.

What's in this chapter?

Ups, downs, a buffet, an open bar, and all the characters from this book. We will be there as well, but please don't bother us too much—we're at work.

Be warned . . . there is an assumption that marriage, like math education, will work for everyone. Please feel free to develop your own unique relationship with the institution, and please, do try the *Tuna Tartar Tacos with Chipotle Aioli*; they are . . . stunning.

#1. Find the Tables

Solve the equation next to each wedding guest's name to see which table (#1 through #6) they will be sitting at for Bart and Sonya's wedding reception.

Mr. Ball, *best man*	___ is the loneliest number
Anita, *mother of the bride*	$x^2 + 3x + 42 = 46$
Natalya, *maid of honor*	$14x + 7 = 21$
Pascal, *Natalya's +1*	$6x - 2 = 28$
Mr. Shaft, *groomsman*	$16 \div 8$
Norah, *stepmother of the groom*	$5^x + 8x + 30 = 71$
Cody, *half brother of the groom*	It takes x to make a thing go right
Lea, *friend of the groom*	Mambo #___
Kurt, *father of the groom*	$\lim_{x \to \infty} \frac{x^2 + 1}{x^2}$
Sammy, Ashley, Andrea, *sisters of the bride*	$\sqrt{9}$
Andy, *Lea's +1*	$2 + (-3) + 6 = x$
McNiel, *your ~~rival~~ best bud*	(# of US states) − 46
Kevin Bacon, *movie star*	$3x^2 + 1 = 109$
You, *McEvoy*	$10 - 3 + x = 11$
Taylor Marie and Mike, *authors*	$4^2 - 10$
Mr. Houdini Prescriptions, *pot dealer*	$x^2 - 4 = 32$

#2. This Is Not A Simulation

Cody's parents haven't been speaking. He's on crutches and can't easily walk away. It's all very unsettling. He's not interested in the amusement his brother rented for the children anyway. It's just a crummy clown and a choo-choo train you can ride in a circle or whatever. He's too old for that kind of nonsense.

Cody looks at Kurt and Norah's stone faces. They remind him of the Dark Castle Simulation. In that one, he was stopped at a crossroads with two guards, one who told the truth and one who was a liar . . .

The wedding is about to start, so Cody has time to ask only one question. Unfortunately, he doesn't know which of his parents is the truth teller and which one is the liar, and he doesn't have his sensory deprivation head mask to pretend this isn't happening.

If Cody wants to find out if his parents are getting divorced, what question should he ask, and of whom?

#3. Vows

SONYA'S VOWS	BARTHOLOMEW'S VOWS
Today I choose you, Bartholomew Midas, to be my partner til death do us part. In sickness and in health, through good times and bad, I promise to always be there for you.	*Sonya LeShine, from the day I met you and gave you that pencil, I knew that you were special. I could tell—I'm really good at stuff like that. Fast forward to today, and hey, what do you know, I was right.*
You are so wise, and strong, and better than those other guys at work. How do you do it?	*You are so funny, and pretty, and smart and funny. How do you do it?*
Bart, I want to spend the rest of my life with you, I promise this in the name of the particular God my parents raised me to believe in, in front of all my friends, and in this dress I'll never wear again.	*Sonya, I want to spend the rest of my life with you, no matter how long that is. A couple days or thousands of years. I give you this ring, a circle, as a symbol of how you make me feel—complete.*
See, I realized, all of this means nothing. Even this beautiful ring means "nothing." Its shape symbolizes what my life would mean without you: zero.	*See, I realized, when I heard you scream and I dropped the $20,000 ring into the mouth of the hellbeast, that it's you that makes me the richest man in the world.*
Thank you, Bart, for your love, which turns my nothing into everything. You've made my dream come true. My dream of not dying alone.	*Thank you Sonya, for your love, which has filled the hole inside me, the hole money could never fill. Thank you for making me whole.*

Draw a picture of Bart and Sonya's rings next to each other, touching. What number does that "equal"?

FACTS & OPINIONS (FROM MIKE)
MONOGAMY

Monogamy is a type of relationship in which a person commits to only one partner.

The word derives from the prefix *mono* (meaning "one") and the suffix *gamy* (meaning "marriage" or "union").

When it comes to the prehistoric origins of monogamy, one theory suggests that males would stay with their mates in order to protect their children from jealous rivals. You see, monogamy is not just about the partners—it's also about the family and the greater community.

Monogamy has many benefits for society:

- Strengthening the family unit
- Reducing the spread of disease
- Encouraging accountability
- Preventing crime (Unmarried men commit a disproportionately high amount of crimes)

Monogamy also has benefits within individual relationships. It builds positive emotions like love, trust, and intimacy, and has even been scientifically linked to longer life expectancies and lower rates of depression.

Some nonhuman animals also engage in monogamous relationships, including wolves, penguins, beavers, gibbons, swans, barn owls, and bald eagles.

Like Math, monogamy is a skill—it takes time and dedication. When you're dedicated to something, you get better at it, and it yields better results. You learn.

In a world where things are changing all the time, there is value in having a companion, someone constant. Someone to praise you at your best moments, accept you at your worst moments, and hopefully, like Math, always try to tell you the truth.

OPINIONS & FACTS (FROM TAYLOR MARIE)
MONOGAMY

Monogamy is a type of relationship in which a person commits to only one partner.

The root of the word is Greek, which should be your first clue that something's not right here.

Actually, your first clue should be *mono*, which means "one." Now, I'm not sitting here saying I'm a mathematician after working on this book, but I'm pretty sure this whole concept is not mathematically sound. How can two people become one? What kind of lesson is that to teach the kids?

Monogamy asks its participants to be the sole providers of contradictory principles:

- A sense of home/adventure
- Mystery/certainty
- Money/happiness (in some cases)

Monogamy as legally binding is even worse. "I love you and it is now the LAW that you and I be with each other and only each other."

Many animals are nonmonogamous, including hyenas, bonobo monkeys, dolphins, hedge sparrows, red elk, bees, and bronze-winged jacanas (an animal worth researching).

Like Math, monogamy is overidealized and sets you up for feelings of failure. But you're not a bad person because you fantasize about your neighbor with the friendly dog. You can't help that their eyes make you feel peaceful.

In a world where the status quo is generally considered to be flawed, there is value in pushing social norms and allowing yourself to explore the relationship that works for you. Whether or not that relationship is monogamous, it shouldn't cost you your freedom as an individual.

#4. The Inner Circle, Part 2

The reception is underway! Everyone gets in line for the alcohol. But Norah's not much of a drinker, so she sneaks away to get high with Kevin Bacon, Mr. Houdini Prescriptions, and us, the authors.

We're all smoking Houdini's exclusive strain, OG Jared Kushner, and feeling just great, like we earned something we didn't work for.

As Norah looks at Kevin Bacon, she has an epiphany: *It's never too late to reinvent yourself! When Kevin Bacon stopped getting lead roles in movies, he reinvented himself as a character actor. I can reinvent myself, too, maybe as a football coach at Cody's school. It'd be nice to share a common interest (if I can get him to care about football). I mean, if he's going to get hurt, I'd like him to get hurt outside.*

Use the following clues to figure out everyone's positions in the circle.

- The authors are not next to each other.
- If Kevin Bacon passes the dutchie on the left-hand side, or the right-hand side, he passes it to a male.
- To the left (to the left, everything you own in a box to the left) of Mr. Houdini Prescriptions are the girls (who run the world? girls).

#5. Cut The Cake

Natalya's fulfilling a maid of honor duty: handing out cake. Anita's with her, looking at photos from about 30 minutes ago.

"Whatever happened to you and Andy? I always thought I'd be at your wedding first."

Natalya looks at the bar, where Pascal is taking shots with Mr. Shaft. Anita looks at her phone, where she's loaded up a website for cat adoption.

"Ooooo, Natalya! Whiskers is looking for a mom."

Natalya can't get a cat, she can't even get a job. No one wants to invest in her company, a nonprofit that turns cigarette butts into school supplies for children.

She wants to talk to Andy, but what will she say? "Hey Andy, good to see you. I didn't reach my potential. I'm a failure. Your new girlfriend sure looks like a free spirit or whatever."

Natalya stares into the cake, which has a radius of 6 wibbles and a height of 4 wibbles. She wants to cut equal pieces for 10 guests, plus one "double piece" for herself. Maybe she'll just put her face in it so she can feel something different.

A) What will be the volume of each piece?

B) What will be the volume of Natalya's double piece?

#6. B-A-See the Reference Section to Solve This One

Pascal is drinking heavily. He takes a whiskey drink, he takes a vodka drink, he takes a lager drink, he takes a cider drink. He weighs 200 pounds. Can he drive home?

#7. What's Mine Is Yours

Bart and Sonya are opening their presents at the wedding, an old Midas family tradition. There are 12 presents sitting on the table. Four of them have blue wrapping, 4 have white wrapping, and 4 have purple wrapping. Eight of them are in boxes, 3 are in bags, and 1 is an envelope addressed "To Whom It May Concern."

To Whom It May Concern:

Congratulations. I, Kurt Midas, president and CEO of the Kurt Midas Oil Corporation, and your father, am pleased to present you with the deed to this farm. Sources tell me there is oil under this land, and I'd like you to own that oil and then sell it to others. After all, oil is forever, and trees are only for now.

I express my congratulations on your nuptials.

Kurt Midas
President and CEO, Kurt Midas Oil Corporation

They open the boxes first, which all turn out to be toasters.

A) What are the chances they'll open a bag next?
B) What are the chances that they'll open the envelope next?
C) What are the chances that Sonya and Bart will ignore Kurt's cries of "Wait! No! Think of the profits!" and sign the farm back over to Anita?

#8. What's My Age Again?

"Well, I know what we're all thinking," Mr. Ball chuckles. "Sonya's just marrying Bart for his money! But let me assure you, his money is my money."

A loud *ZREEEEEUP* is heard. The ballroom doors fly open, and through clouds of neon smoke bursts a homeless man.

"I'VE PREPARED A FEW WORDS FOR BOB AND SARA."

Wait, that's not a homeless man. It's your math teacher, Mr. Newman.

He wipes some blue gunk from his mouth and strolls over to the statue of Nathaniel P. Hogwally, Jr., putting his hand on the statue's shoulder.

"I would just like to say that life, on the whole, is not like Math. Many problems do not have exact solutions. On the other hand, life is just like Math: every problem is an opportunity to learn."

You're a little cheesed. For a guy who made you haul a 6-lb textbook to school every day, this is a pretty simplistic analysis of a complex situation, like something first-time authors would write toward the end of a book.

"Oh, **McEvoy**, **McNiel**, the two best friends from my Period 4 class. How convenient that we are all here, at this wedding. I wanted to tell you something. Space-time works differently and I have aged many years. If you take my age, multiply it by 4, and subtract 2, the answer is 134."

How old is Mr. Newman now?

#9. One Train

Andy just left a conversation among Bart, Lea, and some of their friends from business school. He couldn't appreciate the story about the guy they tricked into putting his life savings into Blockbuster, but maybe you had to be there?

Andy's got his own reminiscing to do. The idea of seeing Natalya today has already brought back memories of their last night in Paris.

"Will you love me forever?" Andy asked.
"I don't know," answered Natalya.
"If you can't give me a certain answer, then what are we doing here?"
"In Paris?"
"No, in general."

He wanders past the clown that was hired for the children. None of the kids are laughing—they're on their phones, cultivating their social media presences.

And there she is. Ten feet away, double piece of cake in her hand, sitting on a miniature train, traveling at a speed of 2 miles per hour. The look on her face is the same as when he saw it last: passionate, questioning, unfulfilled—like an unfinished Sunday crossword.

Natalya turns to see Andy standing on a miniature platform next to a sign that reads: **Must be 36 inches tall to ride**. The look on his face is the same as when she saw it last: gentle, pensive, lost—like a cartoon dog trying to read a map.

Andy joins Natalya on the train. Because it is not built to support two adults, its speed decreases by 50%.

If the track measures 100 feet and the reception ends in 2 hours, how many times can they ride this train together?

#10. Finish Your Homework

It's been a fun day for you and **McNiel**. You met Kevin Bacon, Mr. Newman gave a weird speech, and you popped a 10-second wheelie in the parking lot.

While everyone watches an old man dance with his granddaughter atop his feet, you and your best friend eat the *Spanish-Style Tomato Crostini with Garlic and Olive Oil* and all of the *Brussels Sprouts with Bacon Jam.*

As you rub your tummies, you notice that the rock from page 61 has somehow made its way into the city, crossed the river into this part of town, rolled down Main Street and across six lanes of traffic, up the stairs, through the door, turned left at the bathrooms, past the statue of Nathaniel P. Hogwally, Jr. and made an immediate 90° right into the South Ballroom at the Grand Hypatia Hotel. It lands at your feet.

McNiel grabs a napkin and you pull a pair of scissors out of your backpack. You see now why you didn't cover that rock with fabric. It wasn't because it was too hard, or because you couldn't find the scissors, or because you don't know how to sew. It's because . . . why would you cover a rock with fabric on your own when you could cover a rock with fabric with a friend?

How many people does it take to cover a rock with fabric?

CHAPTER SUMMARY

This is our last chapter summary. It's OK if you want to cry; lots of people cry at weddings. Guys, you can cry, too. Just make sure you do something manly after, like clear your throat.

Let's take another look at what happened here.

1. Bart and Sonya are married now. We have made fun of them a lot, but we wish them the best.

2. Natalya failed to start her organization, a nonprofit that turns cigarette butts into school supplies for children, but she is not a failure. She works hard, is a good friend, and cares about making the world a better place.

3. Cody's parents are getting divorced.

4. Your rival, **McNiel**, is now your best friend.

5. Anita's farm gets to stay in the family.

6. Your math teacher, Mr. Newman, shared some words of wisdom that were not asked for but were appreciated by some.

7. At long last, Natalya and Andy are riding one train.

8. In some ways life reflects Math, but in other ways it does not.

REVIEW QUESTIONS

The South Ballroom of the Grand Hypatia Hotel is empty except for Natalya and Andy, who are still riding the miniature train. They ask the following questions:

1. Natalya → Andy: Can I ask you something that's going to sound weird?

2. Andy → Natalya: How could I possibly say no to that question?

3. Natalya → Andy: Lately, do you feel like . . . you're doing a lot of Math?

4. Andy → Natalya: Wait, yes! Like I never noticed what an integral part of my life it was until recently?

5. Natalya → Andy: Yeah! Like—why do I know that this slice of cake is exactly 24π cubic wibbles?

6. Andy → Natalya: Hey, do you recognize those two people?

7. Natalya → Andy: Which two people?

8. Andy → Natalya: See them, they look high, they're standing near the chocolate fountain?

9. Taylor Marie → Mike: Do you think they can see us?

10. Mike → Taylor Marie: How do we get out of here?

11. Taylor Marie → Mike: Should we just end the story?

BOOK "SUM"MARY

In summation, this book contains:

- 7 chapters
- 57 characters
- 121 problems
- 176 pages
- 29 animals
- 11 mythic creatures
- 1 wedding
- 2 breakups
- 3 trains
- 4 Blink-182 references

And you. **McEvoy**. You were here for it all. Well done, dear friend. Now you can say, "I just finished reading a book about Math" and everyone will give you that look like, *Wow, I'm very impressed with you.*

We do hope you learned something here, whether or not that something was about Math. We hope you laughed. We hope you have a nice day. We hope you'll buy another copy of this book and give it to your brother's girlfriend for her birthday (you don't know her that well).

The book is over. Now it's just you, and whatever you have or haven't learned. Sonya? Bart? Natalya? Andy? They aren't real. I mean, Houdini Prescriptions? That's not a name. **McEvoy** is barely a name.

But the Math was real.

And the way you felt?

That was real, too, dear friend.

That was real, too.

Some Answers

Algebra

#1. Two Trains
See page 28.

#2. What's My Age?
See page 30.

#3. Evaluation
A) 702
B) 1603
C) You'll have to ask her.

#4. Salary-ously, Guys?
See page 33.

#5. Theft!
A) $1000 + 30x = 31x$
$1000 = x$
1000 seconds
B) 16 precious minutes and 40 precious seconds you needed for studying.
C) McNiel thrives on your negative attention, in the same way that Hollywood thrives on scandal.

#6. Quest for the Crucial Information
So *this* is why the teacher made you memorize it.
$2x^2 - 4x - 6 = 0$
$a = 2, b = -4, c = -6$
$x = -1$ or 3

#7. Match Game
1-e, 2-g, 3-f, 4-c, 5-i, 6-b, 7-d, 8-h, 9-Yes, and the answer to this question is one of those things, 10-a, 11-j

#8. You're Like, My Best Friend
Both B and D.

#9. The Space Between

A) Just like problem 4, set the equations equal to each other and solve for the unknown. Seems like you're doing a lot of assuming that everything will equal each other. Is that logical?

$f(x) = g(x)$
$x^2 + 2 = x^2 - 1$
$2 = -1$

Uh-oh, the unknown is gone. This means there is "no solution," which means that the graph has no intersection point.

(This problem exemplifies one of Math's most famous concepts: doing a lot of hard work for nothing.)

#10. Which Came First: The Chicken or the Systematic Pollution of Our Natural Resources?
See page 41.

#11. Mr. Houdini Prescriptions
Let x be the amount of Kevin Wake'n'Bacon sold.
Let y be the amount of Sativa Gomez sold.

Use the following two equations:

$x + y = 15$	$30x + 55y = 650$
$x = 15 - y$	
	$30(15 - y) + 55y = 650$
	$450 + 25y = 650$
	$450 - 30y + 55y = 650$
	$25y = 200$
$x = 15 - 8$	$y = 8$
$x = 7$	

7 Kevin Wake'n'Bacon and 8 Sativa Gomez
(If you're stoned right now . . . good work, it's hard to do Math when you're high.)

#12. Pascal's Numbers

A) Please email a photo of your shape to twotrainsleaveparis@gmail.com.

B) It's 1 (better). Just admit it.

#13. Cookies

See page 45.

#14. Happy Pond, USA

50 cases of chronic respiratory disease this year? That sure is sad.
Set up a proportion:

$$\frac{14}{50} = \frac{365}{x}$$

x is the unknown number of expected cases.

Now cross-multiply, like the bacteria in the water:
$14x = (50)(365)$
$14x = 18{,}250$

Then divide, so that x is alone, like the grieving loved ones of the victims:
$x = 1303.57$

That's a lot of cases! At least doing the math makes it more digestible, unlike the fish in the Grandma River.

Geometry

#1. Square One

See page 51.

#2. Andy Builds an Invisible Fence

$A = b \times h$
$180 = 15 \times h$
$12 = h$
12 feet

#3. Your Chance to Draw (Now That Funding for the Arts Has Been Cut)

A) 

C) 

Draw any shape with thirteen sides.

B) 

D) 

Impossible! Truly perfect circles do not exist in nature, but the sun is the closest. Thank you, Sun, thank you every day for not killing us dead.

#4. True and False: Shapes

1-False, 2-True, 3-True, 4-True, but it is not a *great* song, 5-False, but a square is a type of rhombus, 6-False, they are the same shape but different sizes, 7-True, 8-False, 9-How should we know? 10-False, 11-True

#5. Ninth-Price '90s Media

See page 56.

#6. NSFW

See page 58.

#7. Life Isn't a Highway

$C = 2\pi r$

$20 = 2\pi r$

$\frac{20}{2\pi} = r \approx 3.183$ unpleasant memories

#8. The Box

$V_{\text{big box}} = l \times h \times w$

$V_{\text{big box}} = 30 \times 10 \times 30$

$V_{\text{big box}} = 9{,}000$

$V_{\text{small box}} = l \times h \times w$

$V_{\text{small box}} = 3 \times 1 \times 3$

$V_{\text{small box}} = 9$

$\frac{V_{\text{big box}}}{V_{\text{small box}}} = \frac{9{,}000}{9} = 1{,}000$ boxes

#9. The Rock Show

A) $SA = 4\pi r^2$

$SA = 4\pi(1)^2$ (Use 1, not 2, because the radius is half the diameter)

$SA = 4\pi$

$SA \approx 12.57$ square inches

Make precise cuts, don't want to waste fabric.

B) $A = \pi r^2$

$A = \pi(1)^2$

$A = \pi \approx 3.14$ square inches

#10. The Offer

$A = bh$

$A = (3{,}000)(4{,}000)$ $A = 12{,}000{,}000$ square feet

$12{,}000{,}000 \times \$.10 = \$1{,}200{,}000$

#11. New Apartment

See page 66.

#12. The Inner Circle

Subtract the area of the smaller circle from the area of the larger circle:

$\pi(6)^2 - \pi(2)^2$

$36\pi - 4\pi$

$32\pi \approx 100.53$ square wibbles

#13. Dog-Gone It

A) The volume of coffin A is

$$V = \frac{(24 + 18)(38)(16)}{2}$$

$V = 12{,}768$ cubic inches

So the cost is $12{,}768 \times .08 = \$1{,}021.44$

So coffin B ($999.99) is the better deal, but it's close.

#14. Quest for Salvation

See page 70.

Trigonometry

#1. Find Sine, Cosine, and Tangent
See page 76.

#2. Funding
See page 78.

#3. Periods
7 pregnancy tests when $x = -2.25, -1, -0.25, 1.75, 2.75, 4.25$, and 5

#4. The Perfect Angle
Andy can use the 60° angle, which is acute.

#5. See Your Rival in a New Light
$\tan(20°) = \frac{x}{200}$

$.364 = \frac{x}{200}$

$x \approx 73$ ft

#6. Quarterback in the Day
A) $\sin(50) = \frac{35}{x}$

$x = \frac{35}{\sin(50)}$

$x \approx 46$ yards

B) She *could* wear lingerie and play in the Legends Football League. But in general, men don't like it when women display aggression that can't be sexualized. The systematic exclusion of women from professional sports has a long and storied history. Look up Jackie Mitchell for further reading and look up the original name of the Legends Football League.

#7. Falling Apart
$\sin(47°) = \frac{18}{x}$

$.731 = \frac{18}{x}$

$.731x = 18$

$x \approx 25$ feet

#8. Off the Radar

A) Since right angles are 90° and triangles must add up to 180°, the missing angle must also be 45°, which means the two smaller angles of the triangle are the same measure. Therefore, the two shorter sides of the triangle must be the same length, so Cody can find both of them if he calculates one.

$$\cos(45°) = \frac{x}{1000}$$

$$.7071 = \frac{x}{1000}$$

$$x \approx 707.1$$

He needs to travel 707.1 feet to see the sirens, and 707.1 feet to get back to the port. That's 1414.2 feet in total, which is 414.2 extra feet.

B) The game creators are using all this fun, new technology to enforce scary old narratives.

#9. Bermuda Triangle
See page 87.

#10. What's Your Sine?
Did you think your horoscope would give you a direct answer? Sorry, dear friend, the stars are full of mysteries.

#11. Not That Kind of Love Triangle
They're both right (unlike the angles in this triangle).

Calculus

#1. Ups and Downs
See page 101.

#2. Know Your Limits
See page 103.

#3. Find the Subliminal Messages

Sex. Hey You. Sex. Drink Coke. Sex. Boobs. Your friends only hang out with you (Jesus saves) because they feel sorry for you. Aliens are real and they are coming. No no no. Mom is disappointed in you. Look thin. The government knows about the porn. Aliens.

#4. Moonstruck

Hey, these kids haven't been encouraged to interact directly with other humans. They've learned plenty about sitting in front of screens, which might make it easier on their parents when they get home from camp, but at what cost?

If you want to take the time and do this math, then take the derivative of the function:

$f'(t) = 68 - 1.66t$
$f'(5) = 68 - 1.66(5)$
$f'(5) = 59.7$
Nice (virtual) shot!

#5. Please, Have A Seat

Natalya can sit wherever she wants, or leave and sit outside for a while, or just sit with her feelings. Maybe she'll go sit in a car, on a train, or near a body of water. If she isn't ready to interact with Andy, she doesn't have to, plus, *Space Heroes vs. The Government* is going to be really similar to last summer's *Space Heroes: The Fight for Change.*

Of course, there is the always the option of ignoring all your feelings and doing Math instead:

Let x equal the distance between Natalya and the screen.

Let $f(x)$ equal the viewing angle. Think of the image as two right triangles one inside the other. The viewing angle is going to be the difference between two angles both originating at Natalya, so:

$$f(x) = \cot^{-1}\left(\frac{x}{30}\right) - \cot^{-1}\left(\frac{x}{10}\right)$$

Take the derivative:

$$f'(x) = \frac{1}{10 + \frac{x^2}{10}} - \frac{1}{30 + \frac{x^2}{30}}$$

Set this equal to zero and solve:
$x \approx 17.32$ feet

So she could sit in the 5th row, and lean back.

Or in the 6th row, and lean forward.

#6. Excuse Me, I Need to Take This Call

This one is simple: If you can't be responsible for keeping one oil well from destroying millions of life-forms, you have absolutely no business opening another. Plus, if you spend all your time hanging around people who are just like you, you'll never learn a thing. But hey, at least you'll never be wrong.
The math, however, is complex:

Start with the equation for the area of a circle:
$A = \pi r^2$

Take the derivative of each side with respect to time:

$$\frac{dA}{dt} = 2\pi r \frac{dr}{dt}$$

Plug in the known values and solve for the unknown:

$$10 = 2\pi(528{,}000)\left(\frac{dr}{dt}\right)$$

$$\frac{10}{2\pi(528{,}000)} = \frac{dr}{dt}$$

$$\frac{dr}{dt} \approx .00000301 \text{ ft/sec}$$

Of course, the time it takes to solve this problem could have been spent trying to stop the oil spill from spreading.

#7. An Apple a Day Is Not Sustainable in this Economy

If you want to use Math to calculate how to maximize the apple production, that's your prerogative, but it might take you the same amount of time to just go out and plant a tree.

The apples-per-tree function can be calculated as: $A(x) = -2x + 260$

Which makes the production function: $P(x) = -2x^2 + 260x$

The graph of this is a parabola that opens down, like a frown. Therefore, it will have one absolute maximum, located at its vertex. According to Calculus, the derivative at a graph's vertex is equal to 0.

So, find the derivative: $P'(x) = -4x + 260$

Set it equal to 0 and solve: $0 = -4x + 260$

$$65 = x$$

It's 65 trees she needs. So she shouldn't plant any trees. She should actually go out and cut down 35. That'll be fun.

#8. Edible Arrangements

A-Cinnamon Roll, B-Bowl for Chips and Dip, C-Ice Cream Cone, D-Carnitas Taco

#9. 'Til Death Do We Drop

A) $\frac{\$80{,}000}{12\ months} = \frac{x}{3\ months}$

$12x = 240{,}000$

$x = \$20{,}000$

B) This derivative can be determined visually. At the moment of proposal, the tangent line is horizontal. This means its slope is zero, which means the derivative is equal to zero.

Probability & Statistics

#1. In the Golden Age of Television
See page 121.

#2. Mind the Gap
Oh, you can't calculate that? Very convenient for you.

#3. Dress for Success
See page 123.

#4. Butts

A) $\frac{60}{300} = 1/5 \approx 20\%$

B) $\frac{100}{300} = 1/3 \approx 33.3\%$

C) $\frac{30}{300} = 1/10 \approx 10\%$

D) The chances are pretty small, but everything is impossible until someone tries it.

#5. Shoot
A) Thoughts and prayers go out to the victims and their families.
B) Thoughts and prayers.
C) Thoughts and prayers.
D) Thoughts and prayers.
E) Thoughts.
F) And prayers.
G) Probably high, wouldn't you say? Take a look at the way women's bodies are fully regulated by lawmakers.

#6. Chateau Hellaux
See page 127.

#7. Numerology

Probably not, you spent the whole day thinking about yourself.
(Our numbers are 13.6 [TMF] and 53.9 [MW]. What are yours?)

#8. Aliens Exist

See page 132.

#9. Seating Chart

A) Approach this like Andy's outfit problem. For the first seat, they have 42 options. For the second, 41 options. Then 40, etc. Use your calculator:
42 × 41 × 40 × 39 × 38 × . . . . × 2 × 1 = 1.4050061177529E51
which is approximately 1,405,006,117,752,880,000,000,000,000,000,000,000, 000,000,000,000,000 different ways. Big number.
They'll have to use strategy—put people next to each other who might get along well, and seat Mr. Shaft far from the bar.

B) $3,360.

C) They don't *need* to invite anyone to the wedding. They *choose* to, because they want their loved ones to bear witness to the most important moment of their lives, and because they want to follow the great American tradition of calling attention to yourself.

#10. Piece of the Pie

A) .354 × $1,200,000 = $424,800

B) .104 × $1,200,000 = $124,800

C) .396 × $1,200,000 = $475,200

D) Nothing.

E) (.054 + .473) × $1,200,000 = $632,400
(Note: For Kurt Midas, "donating" also includes political lobbying.)

F) Nothing.

G) Sorry, cannot calculate . . .

#11. Deer Population

A) 26

B) 1 ☹

C) Plug in the current year for *x* in the equation, calculate, and then subtract 1. This deer might get sick and die.

#12. Referendum 18.C.2.D

A) (46,372)(3,000,000)(.003) = \$417,348,000

B) Oh, he can't, you see. He's got to protect his *investments,* and the upkeep on the golf course in Sun Valley is more than you'd expect.

#13. This One Is About You and McNiel

A) $\frac{8+5+15}{3} \approx 9.3$ points per week. Good work!

B)

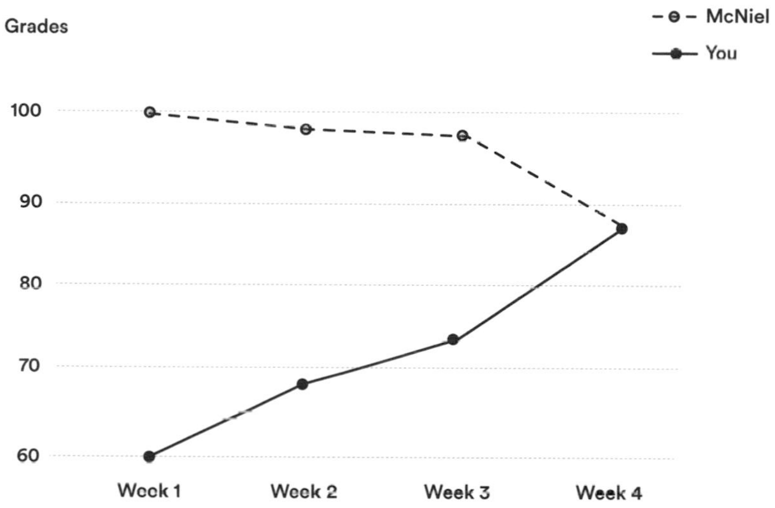

C) If you do this, feel free to cc (or bcc) us at twotrainsleaveparis@gmail.com.

#14. I Crumble for Ya

$.50 \times .40 \times .70 \times .20 \times .36 = .01008 \approx 1\%$

1% chance of Pascal being a good wedding date.

#15. The Grand Hypatia

We're definitely going. There's going to be a bunch of cool stuff there; we heard there will even be amusement for the children.

The Wedding

#1. Find the Tables

Table 1: Mr. Ball, Anita, Natalya, Kurt
Table 2: Mr. Shaft, Norah, Cody
Table 3: Sammy, Ashley, Andrea
Table 4: McNiel, You
Table 5: Pascal, Lea, Andy
Table 6: Kevin Bacon, Taylor Marie, Mike, Mr. Houdini Prescriptions

#2. This Is Not A Simulation

He should ask one of the following:
"Dad, if I asked Mom if you're getting a divorce, what would she say?"
or "Mom, if I asked Dad if you're getting a divorce, what would he say?"

If he asks the truth teller, he learns the false answer, because the truth teller tells Cody what the liar would have said.

If he asks the liar, he also learns the false answer, because the liar lies, telling Cody what the truth teller would *not* have said.

So in either case, if Cody hears, "Son, we are not getting a divorce," he knows for sure that actually, they are.

#3. Vows

∞

Fun fact—on any given night at Smarter Child, that new bar downtown with the arcade games, at least three patrons have a tattoo of this symbol.

#4. The Inner Circle, Part 2

#5. Cut the Cake

A) $V = \pi r^2 h$

$V = \pi(6)^2 (4)$

$V = 144\pi \approx 452.4$

You can consider Natalya's piece to be 2 pieces, so just divide the volume by 12: $144\pi \div 12 = 12\pi \approx 37.7$ cubic wibbles

B) Double the previous answer. Natalya's piece is 75.4 cubic wibbles.

#6. B-A-See the Reference Section to Solve This One

Better not.

#7. What's Mine Is Yours

A) $\frac{3}{4} = 75\%$

B) $\frac{1}{4} = 25\%$

C) 100%.

#8. What's My Age Again?

$4x - 2 = 134$

$4x = 136$

$x = 34$

#9. One Train

$2 - (.50)2 = 1$ mph

At 1 mile per hour, they can travel 2 miles in 2 hours.

That's 10,560 feet.

$10{,}560 \div 100 = 105.6$ times around the track.

#10. Finish Your Homework

It takes at least 2 people to cover a rock with fabric.

THANK YOU

To all of our teachers and students, past, present, and future.

To Samantha Weiner, JL Stermer, Sekka Scher, Patrick Torres & Devin Grosz, who believed in this idea, and then made it real.

To our blood relatives: Leonard Evens, Martha Evens, Sam Evens, Sarah Evens, Cindy Frey, Deacon Frey, Glenn Frey, Otis Frey, Steve Spear, David Wesolowski & Naomi Leah Wesolowski. We love you.

To Jackson Bishop: who listened to the sweet sweet sounds of Math most days while he ate breakfast, sent emails, tried to take naps and showers, and live his own life to the fullest.

To Dallas Spiders Club, who are not from Dallas, nor are they spiders or club members, but the funniest people we have the pleasure of knowing: Sean Brannon, Justin Danforth, Michael Galligan, Paige Goodloe, Shelby LeShine, Lizzie Stewart & Igmalia.

To others, without whom this book would not exist in the way it does now: Jan & Ray Brannon, Bo Burnham, Mike Cheslik, James Coulter, Mary Greer, Jan Hasselbring, Annabel Henley, Emily Levine, Shawn Martin, Mrs. Nelson (who once said: "I used to hate math too, you never know what might happen."), Eden Olsen, Rachel Pollack, Peter Rosemeyer, Bianca Rusu, Matt Schulman, Renna Sheso, Andy Swedberg, Ani Tomasic, Kurt Vonnegut Jr., and, of course, Blink-182.

NO ANIMALS WERE HARMED IN THE WRITING OF THIS BOOK

Editor: Samantha Weiner
Designer: Devin Grosz
Production Manager: Rebecca Westall

Library of Congress Control Number: 2018936283

ISBN: 978-1-4197-3274-4
eISBN: 978-1-68335-495-6

Printed and bound in China
10 9 8 7 6 5 4 3 2 1

ABRAMS The Art of Books
195 Broadway, New York, NY 10007
abramsbooks.com